Smart Materials in Physical and Biological Applications

Dr. Nachieketa K. Sharma

Siksha 'O' Anusandhan University
Bhubaneswar 751030, Odisha, India

United Scholars Publications

United Scholars Publications, USA
Our distinguished editorial team thrives to produce and publish superior quality books and journals.
Visit our website for the current list of publications
www.unitedscholars.net

Editor: Dr. Nachieketa K. Sharma
Siksha 'O' Anusandhan University
Bhubaneswar 751030, Odisha, India

Published by United Scholars Publications, USA

www.unitedscholars.net
info@unitedscholars.net

ISBN-13: 978-0692815212

ISBN-10: 069281521X

Disclaimer

The Publisher and the Editor\Author hold no liability for incidental or consequential injuries or damages caused by the information contained in this publication.

UNITED
SCHOLARS
PUBLICATIONS

ABOUT THE EDITOR

Dr. Nachieketa K Sharma has been teaching Physics to both B. Sc (Hons.) and B. Tech students in different colleges of Odisha, India for last thirty years. He is also actively engaged in the cutting edge theoretical research related to human visual optics, especially the development of biomarkers for early detection of eye diseases like retinitis pigmentosa and age related macular degeneration. He has to his credit over 20 research publications in reputed international and national journals, a book chapter titled Compensative response of retinal cones to loss of luminance efficiency under oblique stimulation in Advances in Natural and Biological Sciences: Volume I published by United Scholars Publications, USA. He has also authored a book titled Compensative Retinal Response to Loss of Luminance Efficiency published by LAP LAMBERT Academic Publishing, Germany in 2016 and also co-authored a book on "Internet" in Odia. At present Dr. Sharma is Professor (Asst.) in the Faculty of Engineering and Technology (ITER), Siksha 'O' Anusandhan University, Bhubaneswar, Odisha.

Preface

A Brief Introduction on the book **Smart Materials in Physical and Biological Applications**

Chapter one deals with the *Bacteriorhodopsin (BR) thin film*-based spatial light modulator (SLM) and erbium/ytterbium (Er^{3+}+Yb^{3+}) doped *polymethylmethacrylate* (PMMA) optical waveguides for the study of retinal response. The proposed polymer optical waveguide model can be improvised for use in the simulation of the retina cone mosaic and retinal implants for patients with less or no vision.

Chapter two is about the plant ScopariadulcisLinn showing both antidiabetic and antihypertensive properties. But as the diuretic and antioxidant properties are related to these two properties, ScopariadulcisLinn also shows diuretic and antioxidant activities. It has shown significant reduction in systolic blood pressure on DOCA induced hypertensive rats.

Chapter three deals with the pseudogap-induced superconductivity used to study the specific heat anomaly in cuprates and analysis of the tunneling spectra. In this chapter the band Jahn-Teller distortion (JT) and the charge density wave (CDW) interactions are considered as pseudogaps for the superconductivity in high-Tc system.

Chapter four deals with the *roots of SidaacutaBurm* F. (Malvaceae) whose preliminary phytochemical investigation indicates the presence of alkaloid, phenolic compound tannic, flavonoid, carbohydrates, glycosides, proteins and amino acids. The ethanol extract of Sidaacuta has exhibited potent post coital, anti-implantation, abortion, estrogenic and antiestrogenic activity and reduction in sperm motility.

Chapter five is about *High dielectric oxides* like TiO_2 and $BaTiO_3$ deposited on Si(111) surface prepared using solid state reaction technique and pulsed laser deposition technique with technological applications for device fabrication.

Chapter six deals with the antidiarrhoeal and anti ulcer activity of the *dietary fiber of C. fistula and T. indica* shown to be the handiwork of the soluble dietary fiber which not only successfully controls the ulcer induced by pyloric ligation but also binds the toxins and rehydrated the intestine by holding the water due to high viscosity thereby serving as an antidiarrhoeal agent also.

Chapter seven is about the antifungal activity of the *binuclear macrocyclic complexes* screened with fungal pathogens (*A.niger, A. flavus,* and *T. viridae*) against a standard antifungal agent Fluconazole and thus found to have robust antifungal property.

Dr. Nachieketa K Sharma

Siksha 'O' Anusandhan University
Bhubaneswar 751030, Odisha, India
November, 2016.

Table of Contents

1. BACTERIORHODOPSIN THIN FILM AND POLYMER OPTICAL WAVEGUIDE: SMART MATERIALS FOR STUDY OF RETINAL RESPONSE by Nachieketa K Sharma 1-33

2. ANTI-DIABETIC, ANTIOXIDANT AND ANTIHYPERTENSIVE ACTIVITY OF SCOPARIA DULCIS LINN. by Manas Ranjan Mishra 34-69

3. THE INTERPLAY OF HIDDEN ORDER WITH SUPERCONDUCTIVITY IN HIGH-Tc SUPERCONDUCTORS by Kamal Lochan Mohanta 70-101

4. ANTIFERTILITY ACTIVITY OF SIDA ACUTA BURM. F by Dusmanta Kumar Pradhan 102-124

5. STUDY OF ELECTRONIC PROPERTIES OF TITANIUM DIOXIDE AND BARIUM TITANATE by Rashmi Rekha Mohanta 125-168

6. ANTIDIARRHOEAL ANTI-ULCER ACTIVITY OF DIETARY FIBER by Ashutosh Mishra & Manas Ranjan Mishra 169-200

7. STUDIES ON TRANSITION METAL COMPLEXES WITH MACROCYCLIC LIGANDS by Sujata Karna 201-224

CHAPTER ONE

BACTERIORHODOPSIN THIN FILM AND POLYMER OPTICAL WAVEGUIDE: SMART MATERIALS FOR STUDY OF RETINAL RESPONSE

Nachieketa K Sharma[1*], Sushanta K. Kamilla[1], Kampal Mishra[2], Jatadhari K Sharma[3], Amir Sahu[4], Ch Sangram Patra[5]

[1]Siksha 'O' Anusandhan University, Bhubaneswar 751030, Odisha, India
[2]Near Mayfair Hotel, Bhubaneswar, Odisha, India
[3]Hemalata Foundation, Balangir, Odisha
[4]Centurion University of Technology and Management, Bhubaneswar, Odisha, India
[5]DAV School, Chandrasekharpur, Bhubaneswar, Odisha, India
**nachiketasharma@soauniversity.ac.in*

ABSTRACT

Bacteriorhodopsin (BR) thin film-based spatial light modulator (SLM) and erbium/ytterbium (Er^{3+}+Yb^{3+}) doped polymethylmethacrylate (PMMA) optical waveguides are employed for the study of retinal response. The bacteriorhodopsin (BR)-based SLM generates an interference pattern of varying contrast on the retina thereby eliciting retinal response of different nature. The proposed bacteriorhodopsin thin film based experimental setup is used to study (a) the retinal response to an interference pattern formed on it (b) the all - optical switching action and (c) the effect of switching action on retinal response. In view of the potential of having a commercial bacteriorhodopsin (BR)-based technology in the near future, the properties of BR_{D96N} (already commercially available variant BR films) vis a vis that of others are also presented.

Similarly, by enhancing the refractive index with an increase in the content of Er^{3+} and Er^{3+}+Yb^{3+} in PMMA, a polymer optical waveguide model is proposed to theoretically obtain individual cone-photoreceptor directionalities and fraction of normally-incident light power coupled to a given cone for the fundamental waveguide mode. These theoretical values are compared with the experimental values obtained from the scanning laser ophthalmoscope imaging method. And it is found that true to the expectation, the spot size of the incident beam is related to the diameter of the photoreceptor indicating a strong directional coupling of the individual cone photoreceptors to the focused incident beam. Moreover, when the incident spot size is identical to the fundamental waveguide mode, i.e., for perfect match the theoretical power-coupling attains one hundred percent efficiency in agreement with the experimental results. Due to low cost, high transparency in the visible spectrum and easy fabrication process, the polymer optical waveguide can be of use both in the simulation of the retina cone mosaic and retinal implants for patients with less or no vision.

Keywords: Stiles-Crawford Effect, Interference Pattern, Bacteriorhodopsin, Retinal Response, Cone Photoreceptor, Light Coupling, Polymer Waveguide.

INTRODUCTION

The photodynamic protein complex bacteriorhodopsin (BR) akin to the visual pigment rhodopsin of the human retina is found in the purple membrane of Halobacterium halobium frequenting the salt marshes at high temperatures [1]. It converts light into an electrochemical ion gradient across the membrane [2]. The absence of photo degradation even after a decade of usage makes this material stable, robust and environmental friendly [3]. This much investigated biomaterial has wide applications in Fourier processing [4-6], volumetric optical memories [5-8], biosensors [9,10], photovoltaic cells [11,12], all-optical switching [13,14], spatial light modulation (SLM) [15-18] and protein-based retinal prostheses [19,20]. In particular, the picosecond response, high quantum yield, sensitivity to low light intensities, photochemical stability and broad wavelength response compared to SLM's slow response times, low sensitivity, poor resolution and short lifetimes [3] enhances BR's importance as a spatial light modulator in studying photoreceptor optics. But in order to have a viable BR SLM, it is suggested that [21] first, BR is to be purified by removing carotenoids, ionic species and other impurities; next, it is to be combined by sol-gel processing with BR film electric field orientation process to produce optically clear, BR/sol-gel (BR/SiO2) nanocomposite films that contain highly-oriented BR molecules locked-in-place within a matrix of rigid glass and leading finally to fabrication of a compatible, working BR-SLM. So, BR/SiO2 films will be more transparent than BR films and will show better SLM response due to their higher BR purity, orientation and concentration. Our motivation behind the choice of this biomaterial for the study of photoreceptor optics lies in the recent advances in genetic engineering and directed evolution characterized by diversification, selection and amplification [22] thereby increasing manifold the potential of introducing a commercial BR-based technology in the near future [23-25].

In one of the important discovery of the last century in human visual optics, the Stiles-Crawford effect of the first kind

(SCE I), a narrow light beam entering at the edge of a pupil elicits a retinal response much weaker than a beam passing near the centre [26]. So long the beam is single, the SCE I diminution of effective brightness for oblique incidence is not different for coherent and incoherent light [27]. But the moment another symmetrically placed coherent beam of equal intensity is added to the first coherent beam from the opposite edge of the pupil, no SCE I visibility reduction is observed [28]. This shows that the ability to interfere proactively influences the traditional SCE I visibility by producing a pattern on the retina [29]. Thus, in addition to revisiting the theoretical computations on the basis of two-wave interference and treatment of SCE I directionality as a light coupling to photoreceptor cones, we have proposed in this Letter a BR based experimental setup which unlike a liquid crystal SLM of finite pixel size [30] is expected not to show any deviations from the theoretical computations reached and also hinted at the all -optical switching action and the effect of switching action on retinal response to an interference pattern formed on it.

An interference pattern can be formed on the retina by using a BR/SiO2 phase-only spatial light modulator which generates two Maxwellian point sources [31] at the pupil and projects a given phase change onto the retina. Thus, the traditional way of determining the SCE I directionality parameter using a single Maxwellian point source scanning across the pupil is avoided [26]. This way we introduce directly a wavefront slope in the plane of the retina and create an interference pattern on it provided the two plane light waves coming symmetrically from two opposite edges of the pupil are capable of interference. Thus, by controlling the contrast of the interference pattern linked to the intensity ratio of the two point sources, the effect of phase variation on SCE diminution of visibility is theoretically computed on the basis of two-wave interference as outlined in the methodology.

The Stiles-Crawford effect of the first kind (SCE I) [26] can be expressed mathematically in two different ways: $\eta = 10^{-\rho_{10}r^2}$ and $\eta = e^{-\rho_e r^2}$, where r is the distance in the entrance pupil from the origin of the function and η is the visibility [32]. The coefficients (of directionality that measures the width of the pupil apodization) ρ_e and ρ_{10} are related by $\rho_e = \ln 10\, \rho_{10} = 2.3\rho_{10}$. The means and standard deviations [33] for ρ_{10} are: horizontal meridian, $0.048 \pm 0.013mm^{-2}$; vertical meridian, $0.053 \pm 0.012mm^{-2}$. In spite of its retinal origin SCE I has been regarded as pupil apodization [29,31,34,35].

But the directionality involved in the cone-specific [36] nature of this effect led Toraldo di Francia [37] to suggest treating cones as radar waveguides which like cones exhibit directional sensitivity. The evidence of the retina being a microstructure with its cones working as waveguides of light was reported by Enoch [38-40]. Thus SCE I was viewed as being created by a guiding of light into the cone photoreceptors [41].

Moreover, it has been experimentally observed that the optical fiber properties of a group of cones resemble those of an individual cone [42]. Incorporation of the diffraction properties of light into the waveguide model 43,44] has led to an analytical expression for the directionality coefficient (ρ_e) in agreement with the SCE I results [45] along with an exodus of experimental works like simulation of a human retinal cone using a liquid-filled photonic crystal fiber [46], measurement of individual cone-photoreceptor directionalities employing a scanning laser ophthalmoscope (SLO) [47], enhancement of the numerical aperture of the eye using an ultrasmall spot size scanning laser ophthalmoscope [48], wavefront sensing with large apex-angle axicon [49], use of a current-controlled tunable liquid-polymer lens to have a greater control on the light hitting the retina [50], spatial and spectral examination of specific retinal behaviour with tuneable spectral liquid-crystal filters [51], use of a near-IR scanning laser ophthalmoscope for *in vivo* directional imaging of the retinal cone mosaic [52].

The recent reporting of the successful design, fabrication and measurement of polymethylmethacrylate (PMMA) optical waveguides by modification of surface profile by applying electric field [53] and optical properties of erbium and erbium/ytterbium doped PMMA [54] has motivated us to study human photoreceptor optics with optical polymer waveguides. Of particular interest is the enhancement in the refractive index with an increase in the content of Er^{3+} and Er^{3+}+Yb^{3+} in PMMA. For our theoretical modeling, we have taken 1% Er^{3+}20%Yb^{3+} doped PMMA at 632.8 nm wavelength as core (refractive index of 1.495) and crown glass of refractive index 1.485 as cladding [55]. The preference of a polymer optical waveguide over the silica type or PCF (photonic crystal fiber) [44,56,57] is due to low cost, high transparency in the visible spectrum and easy fabrication process and the potentiality it promises to offer in retinal implants for patients with less or no vision [58] and simulation of the retina cone mosaic [46].

In this chapter, the waveguide coupling and directionality for that waveguide parameter V which supports a fundamental mode is studied using a proposed optical waveguide fabricated out of polymethylmethacrylate (PMMA) and crown glass. Based upon this polymer waveguide model theoretical values are computed for individual cone-photoreceptor directionalities and fraction of normally-incident light power coupled to a given cone for the fundamental waveguide mode and then compared with the similar measurements obtained from the scanning laser ophthalmoscope imaging method [47,48]. Again, the directionality coefficients are predicted at different foveal eccentricities where the radius of the cone is restricted between 1-1.5 μm and are checked with experimental findings [33,59].

METHODOLOGY

Bacteriorhodopsin (BR) Thin Film-Based Spatial Light Modulator (SLM)

When two plane light waves capable of interference and coming symmetrically from two opposite edges of the pupil meet at a point on the retina, an interference pattern of intensity (I) and phase (φ) is formed with a contrast(m) dependent both on the intensity of the individual light waves (I_1, I_2) and the underlying phase difference $[kx-(-kx)=2kx]$ between them where k denotes the magnitude of the incident wavevector component projected onto the retina [60].

$$\frac{I}{I_1+I_2}=1+m\cos(2kx)$$

$$\tan\varphi=-\left[\frac{(m-1)-\sqrt{1-m^2}}{(m+1)+\sqrt{1-m^2}}\right](\tan kx)$$

$$m=\frac{2\sqrt{I_1I_2}}{I_1+I_2} \quad (1)$$

As seen from Eq. (1), for unit contrast ($m=1$) the wavefront slope contained in the phase is zero (for $I_1=I_2$) But, more the intensities of individual beams are unbalanced, more is the departure of contrast from unity leading to a gradual enhancement of wavefront slope at the retina as exhibited in Fig. 1.

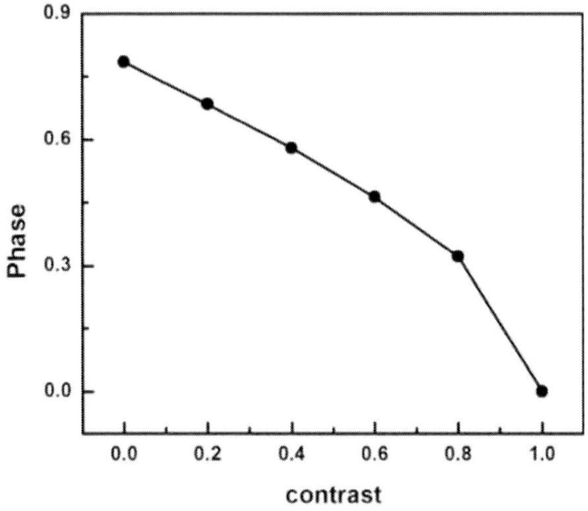

Fig. 1: With the gradual enhancement in the value of contrast (*m*) from 0 to 1 there is a corresponding decrease in the value of the wavefront slope at the retina given by the phase (φ) in radian from Eq. (1) at $x = \frac{\lambda}{8}$

Hence, the enhancement in wavefront slope (φ=0), for axial entry and is $\frac{\pi}{4}$ for peripheral entry for a fixed retinal location, say, $x = \frac{\lambda}{8}$) with the gradual shifting of the light beam's entry into the eye from the centre to the edge of the pupil modifies the pupil entry point (*r*) with contrast as follows [30]

$$r_{eff} = \left[\frac{(m-1) - \sqrt{1-m^2}}{(m+1) + \sqrt{1-m^2}} \right] r \qquad (2)$$

where 'r' is the pupil entrance point from the peak of the visibility.

In the absence of the interference pattern the traditional SCE I visibility (η) is given as [33]

$$\eta(r) = e^{-0.115 r^2} \qquad (3)$$

where the conventional SCE parameter is 0.115/mm^2.The replacement of r with r_{eff} gives the modified visibility (in presence of interference) as

$$\eta(r, m) = e^{-0.115 r_{eff}{}^2} = e^{-0.115\left(\frac{1-m}{1+m}\right) r^2} \qquad (4)$$

The theoretical computations leading to the results and conclusions discussed above can be experimentally verified by adopting a novel procedure of using bacteriorhodopsin thin film as a spatial light modulator as outlined below. The efficiency with which light is coupled to the small-sized photoreceptor cones having low refractive index contrast in the fundamental waveguide mode ultimately depends on the angle of incidence and the waveguide acceptance angle 39,41,43]. Thus, by changing the location and intensity ratio of the two coherent Maxwellian sources a resulting wavefront slope can be generated at the retina leading to the formation of an interference pattern on the retina. So interference pattern of what contrast will be formed on the retina depends on the angle at which light hits the photoreceptors.

In the Fig. 2 described below, first the pump beam is switched off to activate the BR/SiO2 phase-only SLM action by generating holographic phase maps, creating two coherent Maxwellian point sources at the pupil and projecting the phase variation of the SLM onto the retina. A change in the phase profile on the spatial light modulator brings about a change in the contrast of the interference pattern through a variation in the intensity ratio of the two coherent sources. Similarly, the separation between the two sources and the orientation in the pupil plane are governed by grating period on the BR SLM and the set of holographic phase maps generated for each projection angle respectively. Along with the application of BR as SLM we have also proposed in the same experimental setup the realisation of an all-optical switching action by the use of a pump beam of 532 nm to control the transmission of the 632.8 nm He-Ne probe beam and connecting the transmitted power of the probe beam to a computer-controlled optical powermeter [13] not shown in Fig. 2. As the contrast ratio, one of the parameter for evaluating the performance of a simple on-off optical switch varies with pump illumination rate, probe wavelength, and pump and probe beam intensities we can have a control on the switching response of the BR films by varying these parameters. And we can also gain valuable insight as to what happens to the retinal response in terms of the SCE I visibility by simultaneous illumination of the BR sample with pump and probe light beams.

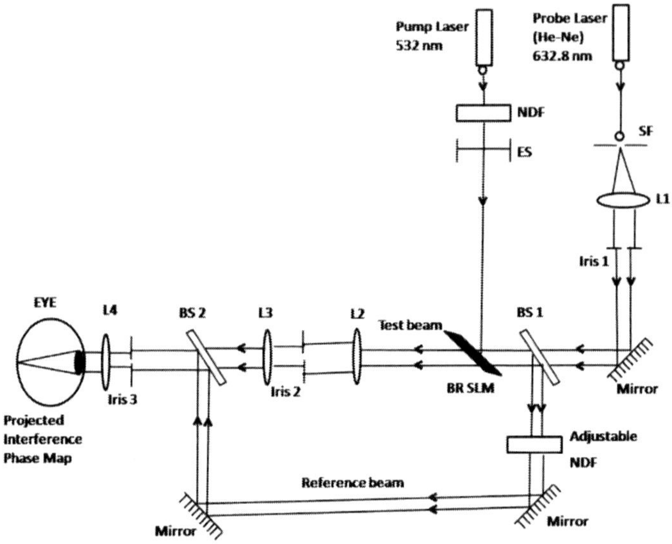

Fig. 2: Schematic of the experimental setup. A BR SLM spatially phase-modulates an expanded collimated beam from a He-Ne laser. This produces two coherent Maxwellian point sources with adjustable relative brightness and separation in the pupil plane. A rapid uploading of phase maps cause the two points to rotate in the pupil plane. The subject compares the perceived visibility of this test field with that of an adjustable reference field from the same laser that enters the eye near the pupil centre. SF: spatial filter with beam expander; BS, beam splitter; NDFs, neutral density filters; ES, electronic shutter.

In view of the potential of having a commercial BR-based technology in the near future, the properties of BR_{D96N} (already commercially available variant BR films) vis a vis that of others are presented below [61]. We see that for fast response time and high spatial resolution a BR SLM would be preferable. However, the lower contrast ratio calls for earlier cleaning of data by thresholding. And though the spatial energy per pixel of 0.2 pJ/pixel is less compared to other SLMs, still it is four times more than the lower limit of 0.05 pJ/pixel of the 'smart pixels' as quantum well-devices that the semiconductor industry afford to produce in its research laboratories till date.

Table 1: Comparison of properties for Spatial Light Modulators made of BR_{D96N} film, Nematic liquid crystals (LC SLM) and ferroelectric liquid crystals (FLC SLM)

Property	BR_{D96N}	*LC SLM*	*FLC SLM*
Spatial resolution(lines/m)	*>5000*	*80*	*130*
Contrast ratio	*1:10*	*1:40*	*1:30*
Response time (μs)	*50*	*4 x 10^4*	*100*
Input intensity (mW/cm^2)	*100*	*0.1*	*0.2*
Switching energy (mJ/cm^2)	*5 x 10^{-3}*	*4 x 10^{-3}*	*2 x 10^{-5}*
Switching energy per pixel(pJ/pixel)	*0.2*	*600*	*1.3*
Electronic power supply	*No*	*Yes*	*Yes*

Erbium/Ytterbium (Er^{3+}+Yb^{3+}) Doped Polymethylmethacrylate (PMMA) and Crown Glass Optical Waveguide

The entire theoretical approach of the model can be outlined as follows: A collimated and spatially filtered He-Ne laser of wavelength 632.8 nm is to be used as a source. This sends out a Gaussian beam whose intensity drops to 14% of its peak value at distance ω_o on either side of the on-axis value. This is then focused on the retina to a spot size of $\omega_r = \frac{\lambda f_{eye}}{\pi n_{eye} \omega_o}$ in the absence of aberrations, where the eye parameters used are those of the reduced eye [62], i.e., a focal length f_{eye}= 22.2 mm and a constant index of refraction n_{eye}=1.33. The focused beam couples its power at the retina to the guided modes of a photoreceptor. And a small number of modes share among them the total power carried forward by the photoreceptor. As the optical polymer waveguide under consideration is cylindrical in nature, the entering beam is Gaussian and when it matches to the location and width of the photoreceptor ($\frac{\omega_r}{\omega_m} = 1$) ($\omega_r$: incident spot size, ω_m: waveguide mode spot size) perfectly, the coupling to the fundamental mode (LP_{01}) becomes the largest. Or else the incident field may couple to higher order modes. The actual number of possible modes is found from the V number of the waveguide defined by $V = \frac{2\pi r_i}{\lambda}\sqrt{n_i^2 - n_s^2}$ where n_i and n_s are the indices of refraction of the inner segment (assumed to be uniform) and the surrounding cladding respectively with r_i being the radius of the photoreceptor cone [63]. For our waveguide and foveal cones, r_i takes values [59] from 1 to 1.4 μm. So only the fundamental mode is allowed [53-55] as the values of V lies between 1.7 to 2.4 with $\lambda = 632.8\, nm$, $n_i = 1.495$ and $n_s = 1.485$. For fundamental mode, V must be less than V_o = 2.405 which happens if the foveal cone is less than 2.218 λ in radius [63].

Again, in our model as the incident beam couples light only to the fundamental mode represented by a Gaussian function of width $2\omega_m$ the fraction of power transmitted to the photoreceptor if incident with its peak value at the photoreceptor axis can be found as [63]

$$T(\theta) = \left[\frac{2\omega_r\omega_m}{\omega_r^2+\omega_m^2}\right]^2 exp\left[\frac{-2(\pi n_{eye}\omega_r\omega_m)^2\theta^2}{\lambda^2(\omega_r^2+\omega_m^2)}\right] =$$

$$\left[\frac{2\omega_r\omega_m}{\omega_r^2+\omega_m^2}\right]^2 exp\left[\frac{-2(\pi n_{eye}\omega_r\omega_m)^2 r^2}{\lambda^2(\omega_r^2+\omega_m^2)f_{eye}^2}\right] \text{or}$$

$$T(r) = \left[\frac{2\frac{\omega_r}{\omega_m}}{1+\left(\frac{\omega_r}{\omega_m}\right)^2}\right]^2 exp\left[-2\left(\frac{\pi n_{eye}}{\lambda f_{eye}}\right)^2\left(\frac{\omega_r^2}{1+\left(\frac{\omega_r}{\omega_m}\right)^2}\right)r^2\right]$$

where $\theta = \frac{r}{f_{eye}}$, $\left(\frac{\omega_r}{\omega_m}\right)$ is the spot size ratio and r is the distance from the on-axis value. Thus, $T(\omega_r/\omega_m) = \left[\frac{2\frac{\omega_r}{\omega_m}}{1+\left(\frac{\omega_r}{\omega_m}\right)^2}\right]^2$ for the beam passing axially through the pupil (i.e., r = 0). Now, the directionality coefficient can be obtained by comparing the fraction of coupled incident light, i.e., $exp\left[-2\left(\frac{\pi n_{eye}}{\lambda f_{eye}}\right)^2\left(\frac{\omega_r^2}{1+\left(\frac{\omega_r}{\omega_m}\right)^2}\right)r^2\right]$ with the SCE I diminution of visibility given by the empirically observed relation[3] i.e., $\eta = e^{-\rho_e \times r^2}$.This leads to:

$$\rho_e(r_i, \omega_r, V) = 2\left(\frac{\pi n_{eye}}{\lambda f_{eye}}\right)^2 \frac{\omega_r^2}{1+\left\{\frac{\omega_r}{r_i\left(0.65+\frac{1.619}{V^{3/2}}+\frac{2.879}{V^6}\right)}\right\}^2}$$

DISCUSSION

Bacteriorhodopsin (BR) Thin Film-Based Spatial Light Modulator (SLM)

Fig. 1 shows that the dependency of the wavefront slope at the retina given by phase (φ) on contrast (m) is crucial in linking the SCE I directionality to a given phase variation which is projected onto the retina through holographic phase maps with BR/SiO_2 phase-only Spatial light modulator. Similarly from Eq. 4 is obtained the dependence of visibility (η) on the contrast (m) in addition to the pupil entry point (r). Changing either k or changing m for a fixed value of k allows tuning of the wavefront slope at the retina. For intermediate situations between the cases of axial (m =1) and most oblique (m =0) incidence, the wavefront is steepest nearest to the intensity minima as indicated in Fig. 3.

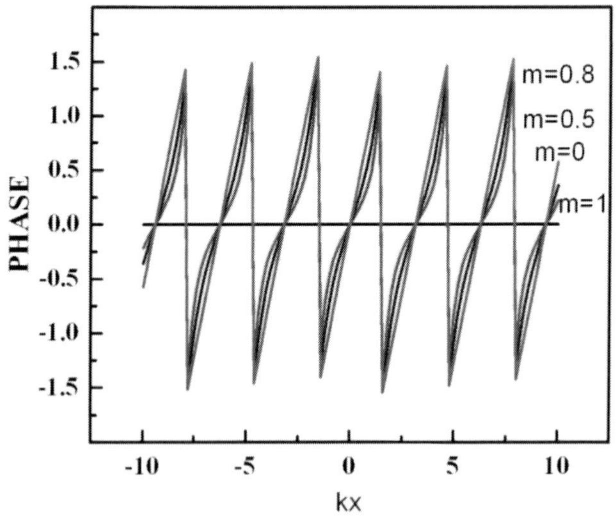

Fig. 3: Phase (φ) in radian is plotted against kx (in radian) for various values of contrast (m). Changing either k or changing m for a fixed value of k allows tuning of the wavefront slope at the retina. For intermediate situations between the cases of axial (m = 1) and most peripheral (m = 0) incidence, the wavefront is steepest nearest to the intensity minima.

The intensity distribution for unit contrast (m = 1), zero contrast (m = 0) and for intermediate values of contrast lying between the axial and most peripheral case (here, m = 0.5 and m = 0.8 are considered) is graphically presented in Fig. 4.

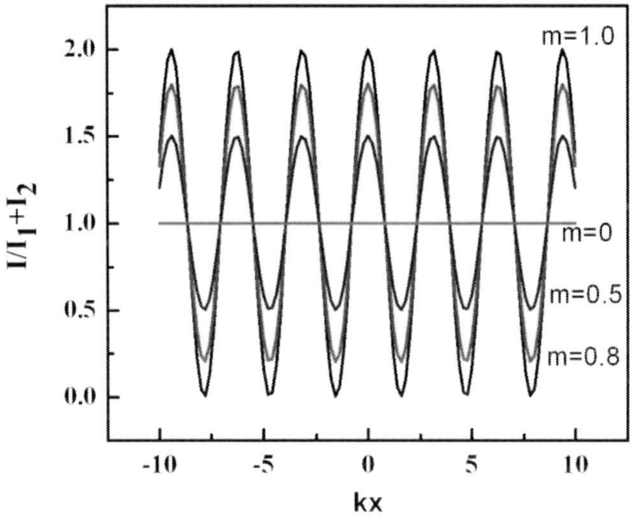

Fig. 4: The intensity distribution (where I_1, I_2 = intensity of individual light waves and I is intensity of the interference pattern) for m = 1, m = 0.5, m = 0.8 and m = 0.

From the above Fig. 4 it is seen that the phase gradient increases at points of constructive interference. Again, Eq. 4 shows the dependence of visibility η *(r, m)* on the contrast (*m*) in addition to the pupil entry point (*r*). The behaviour of visibility in the presence of the twin parameters of contrast and pupil entry point is graphically explored in Fig. 5. When the contrast is unity (*m*=1), the interference pattern does not leave any trace of wavefront slope (φ=0) of light on the retina as revealed from Fig. 1. Behaviour mimics an axial beam with no SCE visibility reduction. Somehow the retina forgets the physical entry point of the beams and rather prefers to respond to the pattern they have left on it. The SCE I is completely switched OFF. But when the contrast is zero or approaches to infinity (m = 0 or $m \rightarrow \infty$), in place of two beams only one (as either I_1 or I_2 = 0) is incident on the eye. Then the interference pattern on the retina vanishes. Immediately the retina senses the pupil entry point of the beam and visibility decreases with increasing pupil entrance point just like conventional SCE visibility. The wavefront slope picks up (φ_{max}) with a consequent diminution in visibility (minimum η. The SCE I is switched ON. For intermediate values of contrast between 1 and 0(*m)* = 0.9, 0.8, 0.5) visibility (η) increases with increase of *m* with corresponding decrease in the wavefront slope at the retina.

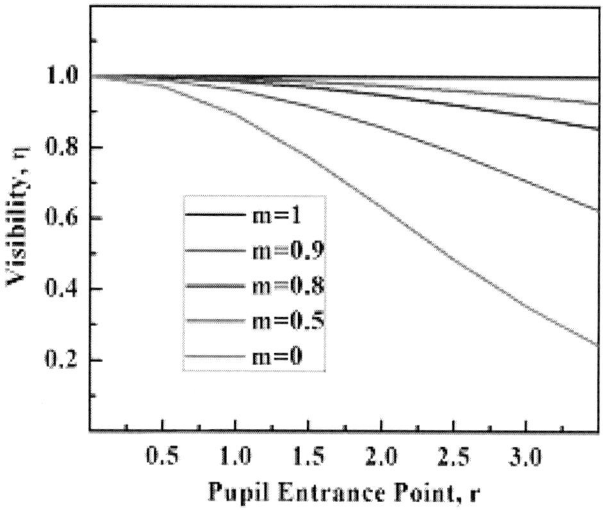

Fig. 5: Variation of visibility (η) with pupil entry point (r) for different contrast (m) of the interference pattern. Again, these visibility curves are obtained for foveal cones having a directionality parameter of 0.115/mm^2

Finally, the sole effect of contrast on visibility for fixed pupil entry point is studied in Fig. 6. As seen from the figure, when the contrast is slowly regulated from m= 1 to m = 0 keeping the pupil location fixed at r= $3.5mm$, the visibility responds accordingly by decreasing to almost 24 % of its maximum value. Thus, the coherence in the entering beam governs the retinal stimulation much more predominantly than the SCE I diminution of effective visibility dependent on the pupil entry point of the beam. When the interference pattern is full blown with m = 1, the SCE I is completely eclipsed, but its control over the retinal stimulation gradually increases in proportion to the corresponding decrease in the value of the contrast from $m = 1$ to $m = 0$. And in the absence of interference pattern from the retina, the SCE I solely controls visibility.

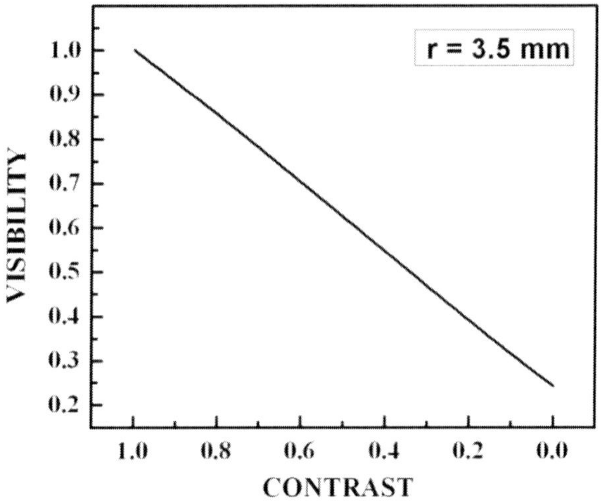

Fig. 6: Visibility (η) is found to decrease with the decrease of the contrast (m) from 1 to 0 in accordance with the corresponding enhancement in the wavefront slope (φ) at the retina for a fixed value of the pupil entry point (r = 3.5 mm).

Erbium/Ytterbium (Er^{3+}+Yb^{3+}) Doped Polymethylmethacrylate (PMMA) and Crown Glass Optical Waveguide

We see from Fig. 7 that light coupling to only fundamental mode is allowed as V is less than $V_o = 2.405$ which happens for cones having radii less than 2.218λ (or 1.4 μm) [63].

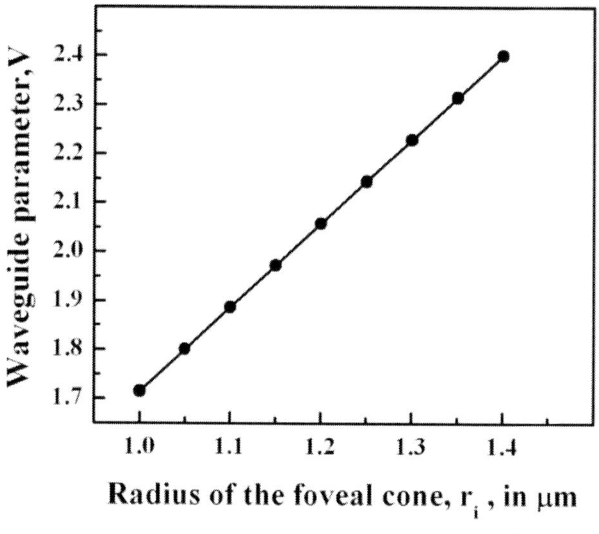

Fig. 7: Variation of waveguide parameter, V with the radius of the photoreceptor cones at the foveal site

Thus, at the foveal site, $V < V_o = 2.405$ [63] allows the incident beam to couple light only to the fundamental mode represented by a Gaussian function of width $2\omega_m$, where ω_m is the waveguide mode spot size. For a single-mode condition the mode-width is related to the radius of the waveguide (that is, the photoreceptor radius, r_i) given by the empirical relation [64] $\omega_m = r_i\left(0.65 + \frac{1.619}{V^{\frac{3}{2}}} + \frac{2.879}{V^6}\right)$. The variation of waveguide mode spot size (ω_m in μm) with the radius of the individual photoreceptor cone (also in μm)at the foveal site for a fixed waveguide parameter V = 2.0 is shown in Fig.8.

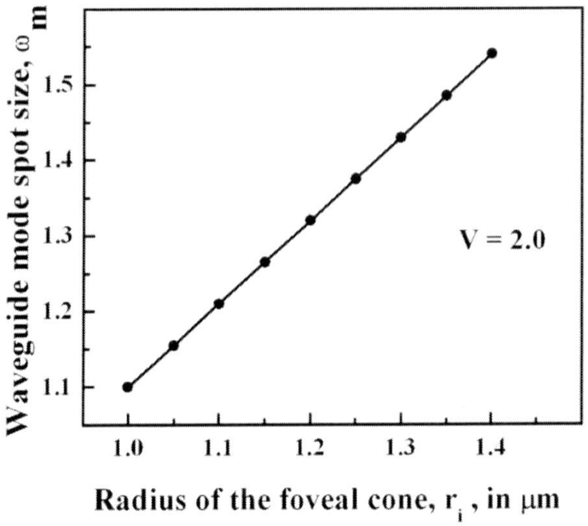

Fig. 8: Response of waveguide mode spot size ($\omega_m in\ \mu m$) to the radius of the photoreceptor cones at the foveal site

And how the waveguide mode spot size responds to the waveguide parameter V for a particular foveal photoreceptor cone is illustrated in Fig.9.

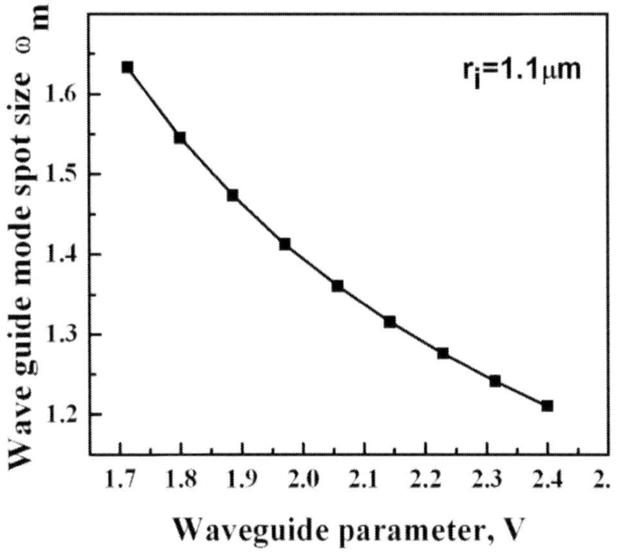

Fig. 9: Variation of waveguide mode spot size (ω_m in μm) with the waveguide parameter, V

This coupled power T is plotted against spot size ratio in Fig.10 and we got the result that for perfect matching, i.e., $\frac{\omega_r}{\omega_m} = 1$ it attains one hundred percent efficiency. But when the spot size ratio departs from the perfect match either decreasing by a factor of half or increasing by a factor of two, the coupled power responds by dropping to 64 % of its peak value. Similarly, a mismatch of one- fourth or four times i.e., when $\omega_m = \frac{\omega_r}{4}$ or $\omega_m = 4\omega_r$, the light coupling registers an identical weakening by becoming 22 % of the maximum coupled efficiency for perfect match. This is beautifully depicted in Fig. 10 and verified experimentally [47,48].

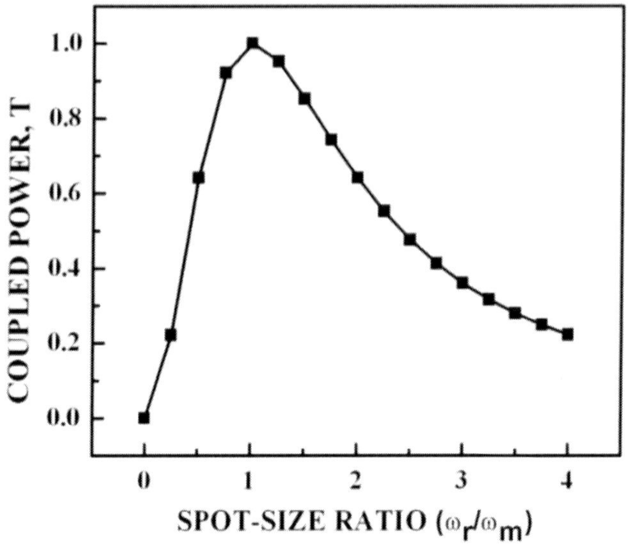

Fig. 10: Response of power coupled to the fundamental waveguide mode to spot size ratio. At perfect matching condition, i.e., $\omega_m = \omega_r$ (waveguide mode spot size equals the incident spot size) the coupled power attains 100% efficiency.

The variation of directionality coefficient ρ_e (in mm^{-2}) against photoreceptor radius r_i (in μm) is plotted in Fig. 11 for various incident spot size (ω_r= 1.0, 1.1, 1.2, 1.3, 1.4, & 1.5μm) at fixed waveguide parameter of V = 2.0 which supports the fundamental mode. For f_{eye}=22.2mm, n_{eye}=1.33, r_i= 1.2μm, λ=632.8nm, ω_r =1.0μm, V=2.0, ρ_e=0.1236/mm^2 or ρ_{10} =0.053mm^2which agrees with the observed SCE I directionality parameter [33] ρ_{10} = 0.053±0.012mm^{-2}. When the beam is well matched to the fundamental mode of the photoreceptor, the directionality parameter, ρ_e can be obtained as $\rho_e = \left(\frac{\pi n_{eye}\omega_r}{\lambda f_{eye}}\right)^2$

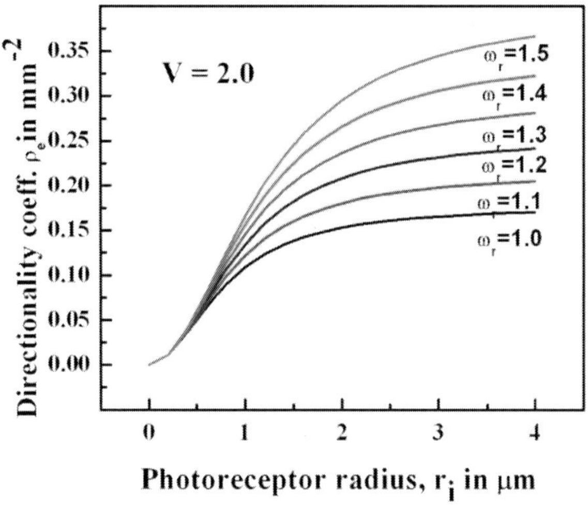

Fig. 11: Variation of directional coefficient to the photoreceptor radius for different incident spot size. For a fixed radius, directionality increases with increase of ω_r indicating a strong directional coupling.

As revealed from Fig.11 the directionality is highly dependent on the light coupling mechanism of the incident beam to the photoreceptor, in this case particularly manifested in the relation between the spot size of the incident beam and the radius of the photoreceptor as confirmed by the use of scanning laser ophthalmoscope [47,48].

CONCLUSION

So we can summarise that the retinal response to coherent illumination can be studied by generating an interference pattern of varying contrast on the retina employing a smart material in the form of a Bacteriorhodopsin (BR) thin film-based spatial light modulator (SLM). We can conclude that, with an interference pattern no more present on the retina, the retinal stimulation takes the traditional SCE I route, but with a pattern of unit contrast the SCE I becomes totally irrelevant for the retina. For intermediate situations (between $m = 1$ to $m = 0$) the visibility diminution with increasing pupil entrance point, characteristic of traditional SCE gets an opposing boost from the interference pattern on the retina in commensurate with its contrast.

Similarly, the conclusion of why the retina responds to an interference pattern formed on it the way it is responding as revealed from the employment of a BR film based SLM can be more deeply investigated by use of another novel smart material, that is, Erbium/ytterbium (Er^{3+}+Yb^{3+}) doped polymethylmethacrylate (PMMA) and Crown glass optical waveguides modeled as a photoreceptor cone [65]. And true to the expectation; the spot-size of the incident beam (ω_r) is related to the radius of the photoreceptor (r_i) cone indicating a strong directional coupling of the individual cone photoreceptors to the focused incident beam. And again it is seen that when the incident spot-size is identical to the fundamental waveguide mode the theoretical power-coupling attains one hundred percent efficiency for perfect match in agreement with the experiment. Thus, for a focused beam of light, the directionality turns out to be highly dependent on the ratio between the spot-size of the incident light and the size of the mode propagated along the photoreceptor. The proposed polymer optical waveguide model can be improvised for use in the simulation of the retina cone mosaic and retinal implants for patients with less or no vision.

REFERENCES

[1] D. Oesterhelt and W. Stoeckenius, Rhodopsin-like protein from the purple membrane of Halobacterium halobium, Nature, New Biol. **233**, (1971) 149-152.

[2] P.N. Prasad, Introduction to Biophotonics, Wiley-Interscience, New York (2003).

[3] www.eqbs.ump.edu.pl/presentations/eqbs_rao.pdf

[4] D. Oesterhelt, C. Brauchle and N. Hampp, Bacteriorhodopsin: a biological material for information processing, Quarterly Reviews of Biophysics, 24, (1991) 425-478.

[5] R. R. Birge, N. B. Gillespie, E. W. Izaguirre, A. Kusnetzow, A. F. Lawrence, D. Singh, Q. Wang Song, E. Schmidt, J. A. Stuart, S. Seetharaman, and K. J. Wise, Bimolecular electronics: Protein-based associative processors and volumetric memories. J. Phys. Chem. B, **103** (1999) 10746-10766.

[6] J.A. Greco, N.L.Wagner, M.J. Ranaghan, S. Rajasekaran and R. R. Birge, Bimolecular Information Processing: from Logic Systems to Smart Sensors and Actuators, edited by E Katz, Wiley-VCH, Weinheim, Germany (2012), p. 33.

[7] N. Hampp, Bacteriorhodopsin as a Photochromic Retinal Protein for Optical Memories, Chemical Reviews, 100(5), (2000) 1755-1776.

[8] J. A. Stuart, D.L. Marcy, K. J. Wise and R. R. Birge, Volumetric Optical Memory based on bacteriorhodopsin, Synth. Met. **127** (2002) 3-15.

[9] F. Boucher, S. Taneva, S. Elouatik, M. De´ry, S. Messaoudi, E. Harvey-Girard, and N. Beaudoin, Reversible Inhibition of Proton Release Activity and the Anesthetic-induced Acid-Base Equilibrium between the 480 and 570 nm Forms of Bacteriorhodopsin, Biophys. J. **70**, (1996) 948-961.

[10] J. Lanyi and H. Luecke, Bacteriorhodopsin, Curr.Opin. Struct. Biol., 11(4) (2001) 415-419.

[11] F. T. Hong, Retinal Proteins in Photovoltaic Devices, Adv. Chem. 240, (1994) 527-560.

[12] J. Xu, A. B. Stickrath, P. Bhattacharya, J. Nees, G. Varo, J. R. Hillebrecht, L. Ren and R. R. Birge, Direct Measurement of the Photoelectric Response Time of Bacteriorhodopsin via Electro-Optic Sampling, Biophys. J. 85 (2) (2003) 1128-1134.

[13] R.K. Banyal and B.R. Prasad, High-contrast, all-optical switching in bacteriorhodopsin films, Appl. Opt. 44 (26) (2005) 5497 -503.

[14] Y.Huang, S. T. Wu, and Y. Zhao, All-optical switching characteristics in bacteriorhodopsin and its applications in integrated optics, Opt. Express 12(5) (2004) 895-906.

[15] D. Sanchez, Drew A. Pommet, and Michael A. Fiddy, "Compact optical correlator for machine vision with optically addressed bacteriorhodopsin spatial light modulator," Proc. SPIE 2908, Machine Vision Applications, Architectures, and Systems Integration V, 198-203 (October 31, 1996) Boston, USA

[16] Q. Wang Song, Chungping Zhang, Robert Blumer, Richard B. Gross, Zhongping Chen, and Robert R. Birge, Opt. Lett., 18, (1993) 1373.

[17] F. J. Aranda, R. Garimella, N. F. McCarthy, D. Narayana Rao, D. V. G. L. N. Rao, Z. Chen, J. A. Akkara, D. L. Kaplan, and J. F. Roach, All-optical light modulation in bacteriorhodopsin films, Appl. Phys. Lett., 67(5), (1995) 599-601.

[18] Jesper Gluckstad and P. S. Ramanujam, Lasers and Electro-Optics Society Annual Meeting, 1994. LEOS '94 Conference Proceedings, IEEE, 2 (1994) 318.

[19] N. L. Wagner, J. A. Greco, and R. R. Birge, Bionanotechnology: Biological Self-Assembly and its Applications edited by B. H. A. Rehm, Caister Academic Press, Norfolk, UK, (2013) p. 205–240.

[20] Z. Chen and R. R. Birge, Protein-based artificial retinas, Trends Biotechnol. 11(7), (1993) 292-300.

[21] https://www.sbir.gov/sbirsearch/detail/169239

[22] Nicole L. Wagner, Jordan A. Greco, Matthew J. Ranaghan and Robert R. Birge, Directed evolution of bacteriorhodopsin for applications in bioelectronics, J. R. Soc. Interface 10(84), (2013)20130197 (2013).

[23] K. J. Wise, N. B. Gillespie, J. A. Stuart, M. P. Krebs, R. R. Birge, Optimization of bacteriorhodopsin for bioelectronic devices, Trends Biotechnology, 20(9) (2002) 387-394.

[24] J.R. Hillebrecht, K.J. Wise, J.F. Koscielecki, R.R. Birge, Directed evolution of bacteriorhodopsin for device applications, Method Enzymol., 388 (2004) 333 -347.

[25] J. R. Hillebrecht, J. F. Koscielecki, K. J. Wise, M. P. Krebs, J. A. Stuart, and R. R. Birge, Nanobiotechnology: Bioinspired Devices and Materials of the Future, edited by O. Shoseyov and I. Levy, Humana Press Inc., Totowa, NJ, (2008) p. 121–140.

[26] W.S.Stiles and B.H. Crawford, The luminous efficiency of rays entering the eye pupil at different points, Proc. R. Soc. B, **112** (1933) 428-450

[27] L.J. Bour and J.C.M. Verhoosel, Directional sensitivity of photoreceptors for different degrees of coherence and direction of polarization of the incident light, Vis. Res. **19** (1979) 717- 719.

[28] B. Vohnsen and D. Rativa, Absence of an integrated Stiles Crawford function for coherent light, Journal of Vision **11**(1) (2011) 1-10.

[29] G.Westheimer, Retinal light distributions, the Stiles-Crawford effect and apodization, J. Opt. Soc. Am. A, **30** (2013) 1417-1421

[30] S. Castillo and B. Vohnsen, Exploring the SCE I with coherent light and dual Maxwellian sources, Appl. Opt., **52** (2013) A1-A8

[31] G. Westheimer, Retinal light distribution for circular apertures in Maxwellian view,J. Opt. Soc. Am., **49** (1959) 41-44

[32] D.A. Atchison, A. Joblin, and G. Smith, Influence of Stiles-Crawford effect apodization on spatial visual performance,J. Opt. Soc. Am. A ,**15** (1998) 2545-51

[33] R.A. Applegate and V. Lakshminarayanan, Parametric representation of Stiles-Crawford functions: normal

variation of peak location and directionality, J. Opt. Soc. Am. A, **10** (1993), 1611-1623

[34] H. Metcalf, Stiles-Crawford apodization, J. Opt. Soc. Am., **55** (1965) 72-73

[35] J.P. Carroll, Apodization model of the Stiles-Crawford effect, J. Opt. Soc. Am. A, **70**(1980) 1155-1156

[36] F. Flamant and W.S. Stiles, The directional and spectral sensitivities of retinal rods to adapting fields of different wavelengths, J. Physiol, **107**(1948) 187-202

[37] G. Toraldo di Francia, Retina cones as dielectric antennas, J. Opt. Soc. Am. **39**, (1949) 324.

[38] J.M. Enoch and G.A. Fry, Characteristics of a model retinal receptor studied at microwave frequencies, J. Opt. Soc. Am. **48**, (1958) 899-911.

[39] J.M. Enoch, Optical properties of the retinal receptors, J. Opt. Soc. Am. **53**, (1963) 71-85.

[40] J.M. Enoch, Waveguide modes: Are they present, and what is their possible role in the visual mechanism? J. Opt. Soc. Am. **50**(10), (1960) 1025-1026.

[41] A.W. Snyder and C. Pask, The Stiles-Crawford effect-explanation and consequences, Vision Res. **13** (1973) 1115-1137

[42] A. Roorda, and D. R. Williams, Optical fiber properties of individual human cones, J. Vision **2**(5), (2002) 404-412.

[43] B. Vohnsen, Photoreceptor waveguides and effective retinal image quality, J. Opt. Soc. Am. A., **24**(3), (2007) 597-607.

[44] B. Vohnsen, I. Iglesias and P. Artal, Directional light scanning laser ophthalmoscope, J. Opt. Soc. Am. A. **22**(12), (2005) 2606-2612.

[45] W. S. Stiles, The directional sensitivity of the retina and the spectral sensitivities of the rods and cones, *Proc. R. Soc. B* ***127****,* (1939) *64*–105.

[46] D. Rativa and B. Vohnsen, Simulating human photoreceptor optics using a liquid-filled photonic crystal fiber, Biomed. Opt. Express **2**, (2011) 543-551.

[47] D. Rativa, and B. Vohnsen, Analysis of individual cone-photoreceptor directionality using scanning laser ophthalmoscopy, Biomed.Opt.Express **2**, (2011) 1423-1431.

[48] B. Vohnsen and D. Rativa, Ultrasmall spot size scanning laser ophthalmoscopy, Biomed.Opt.Express **2**, (2011) 1597-1609.

[49] B. Vohnsen, S. Castillo, and D. Rativa, Wavefront sensing with an axicon Opt. Lett. **36**(6), (2011) 846-848 .

[50] B. Lochocki, and B. Vohnsen, Defocus-corrected analysis of the foveal Stiles-Crawford effect of the first kind across the visible spectrum, J. Opt. **15**(12), (2013). 1-9.

[51] B. Lochocki, D. Rativa, and B. Vohnsen, Spatial and spectral characterization of the first and second Stiles-Crawford effects using tuneable liquid-crystal filters, J. M od. Opt., **58**, (2011) 1817-1825.

[52] B. Vohnsen, I. Iglesias, and P. Artal, Directional imaging of the retinal cone mosaic, Opt. Lett. **29**(9) (2004) 968-970.

[53] V. Prajzler, O. Lyutakov, J. Tuma, I. Huttel, J. Spirkova, V. Svorcik, V. Jerabek, Electroscope, Rocnik (2010).

[54] V. Prajzler, V. Jerabek, O. Lyutakov, I. Huttel, J. Spirkova, V. Machovoc, J. Oswald, D. Chvostova, and J. zavadil, Acta Polytecnica, **48**(5), (2008) 14-21.

[55] *n.wikipedia.org/wiki/List_of_refractive_indices*

[56] P. St. J. Russell, Photonic crystal fibers, Science **299**, (2003) 358-362.

[57] J. C. Knight, Photonic crystal fibers, Nature **424**(2003)., 847-851.

[58] E. Zrenner, K. U. Bartz-Schmidt, H. Benav, D. Besch, A. Bruckmann, V. P. Gabel, F. Gekeler, U. Greppmaier, A. Harscher, S. Kibbel, J. Koch, A. Kusneyrik, T. Peters, H. Sachs, A. Stett, P. Szurman, B. Wilhelm, and R. Wilke, Subretinal electronic chips allow blind patients to read letters and combine them to words, Proc. Biol. Sci. **278**(1711) (2011) 1489-1497.

[59] C.A. Curcio, K.R. Sloan, R.E. Kalina and A.E. Hendrickson, Human Photoreceptor Topography, Journal of Comparative Neurology, **292** (1990) 497-523

[60] Nachieketa K. Sharma, Kampal Mishra, Sushanta K. Kamilla, and Jatadhari K. Sharma, "Using Bacteriorhodopsin Thin Film in Regulating Retinal Stimulation of a Human Eye," *Adv. Sci. Lett.* Vol. 20, No.3-4, pp. 705-709, March/April 2014.

[61] C. Brauchle, N. Hampp, and D. Oesterhelt, Molecular Electronics: Properties, Dynamics, and Applications, edited by G. Mahler, V. May, and M. Schreiber, CRC Press, (1996), p. 3

[62] H. H. Emsley, *Visual Optics*, 5th ed., Vol. 1. Hatton, (**1955**).

[63] A. Ghatak, and K. Thyagarajan, Introduction to Fiber Optics (Cambridge, U.K., pp 149-156 (1998)D. Marcuse, *Bell Syst. Tech. J.* 56, 703 **(1977)**.

[64] D. Marcuse, Loss Analysis of Single-Mode Fiber Splices, Bell Syst. Tech. J. **56**, (1977). 703-718.

[65] Nachieketa K. Sharma, Kampal Mishra, Sushanta K. Kamilla, and Jatadhari K. Sharma, "Modelling of a Retinal Cone Using Erbium/Ytterbium Doped Polymethylmethacrylate and Crown Glass," *Adv. Sci. Lett.* Vol. 20, No. 3-4, pp. 788-791, March/April 2014.

CHAPTER TWO

ANTI-DIABETIC, ANTIOXIDANT AND ANTIHYPERTENSIVE ACTIVITY OF *SCOPARIA DULCIS* LINN

Manas Ranjan Mishra

Department of Pharmacognosy, Gayatri College of Pharmacy, Sambalpur-768200, Odisha, India.
mishra053@gmail.com

ABSTRACT

Herbal drugs have great growth potential in the global market. Natural product research continues to explore Indian Traditional Medicines for development of novel drugs. The problem scientists as well as the consumers face is the lack of proper standardisation techniques. We have a great folkloric assistance from Ayurveda and Vaidyas practising traditional medicine. It is the need of the hour to validate their claims by using modern techniques. Some *in vivo* and *in vitro* models for anti-diabetic, antihypertensive and antioxidant studies were carried out here in order to provide simple and effective assessment of the biological screening of botanicals by using different models without going for extensive extraction and purification steps. After drying and size reduction the plant is stored for further studies like phytochemical evaluation. In the preliminary phytochemical screening the presence of the phyto-constituents in the powder of whole plant has been ascertained.

Basing on the extractive value of different extracts the methanolic extract was selected for further studies. In the pharmacological evaluation we have first gone for the acute toxicity study and on that basis we have determined the doses for further studies. We have performed the anti-diabetic study both on in vitro and *in vivo* models and have found positive results for both the models.Further we have performed the antioxidant activity of methanolic extract of *S.dulcis* (MESD) and water extract of *S.dulcis* (WESD) for comparison. This has shown positive result providing substantiation for its anti-diabetic activity as there is a link between the antioxidant and the anti-diabetic activity. Finally, the antihypertensive activity for both *in vitro* and *in vivo* models is performed. The activity is carried out in *in vitro* models on MESD and WESD. As MESD has shown better result, the *in vivo* test is performed for only this extract which has shown significant reduction in systolic blood pressure on DOCA induced hypertensive rats.

Keywords: Ayurveda, phyto-constituents, MESD, WESD, DOCA, *in vitro* and *in vivo*

INTRODUCTION

Scoparia dulcis is a medicinal herb distributed throughout tropical and subtropical regions of India, America, Brazil, West Indies, and Myanmar. It is introduced in India from tropical America [1,2]. *S. dulcis* (Mala-anis) is a ubiquitous weed, a native of tropical America, found in the Philippines, in and about towns throughout the settled areas at low and medium altitudes. According to Dalziel, when chewed, the plant is at first bitter and later sweet. Sweet-broom, a widespread herb in the tropics, is commonly found as a weed near houses. It would go too far to say it was cultivated, but it is at least tolerated. The most common references to it are its use as broom/stems. These are bundled and used to clean the area around and within homes. This practice is very old and has been reflected in the scientific as well as most of the common names. The name *Scoparia dulcis* was applied by Linnaeus in 1753 [3-6].

The plants were collected from local area of Barpali, Dist-Bargarh, Odisha, during November 2009 to January 2010. The plant was authenticated as *Scoparia dulcis* Linn by Botanical Survey of India, Central National Herbarium Howrah, Kolkata, India (Ref.No:CNH/I-I/38/2009/Tech II/137).

Anti-Diabetic Activity

After the recommendations made by WHO on diabetes mellitus [7], the search for safer and more effective hypoglycemic pharmaceuticals has continued to be an important area of research. The search for novel hypoglycemic compounds from medicinal plants has become an important aspect. Keeping in view of the difficulties in testing *in-vivo;* several *in-vitro* tests have been developed in recent years [8].

The number of diabetic patients is rapidly rising in most parts of the world, especially in developing countries such as India, Thailand and Indonesia. In general, the control of blood glucose concentrations near the normal range is mainly based on the use of oral hypoglycaemic/antihyperglycaemic agents and insulin. However, all of these treatments have limited efficacy and are associated with undesirable side effects [9-11], leading to increasing interest in the use of medicinal plants for the alternative management of type 2 diabetes mellitus. The control of postprandial (PP) plasma glucose levels is critical in the early treatment of diabetes mellitus and for reducing chronic vascular complications [12]. PP hyperglycemia is one of the earliest abnormalities of glucose homeostasis associated with type-2 diabetes and is markedly exaggerated in diabetic patients with fasting hyperglycemia. PP hyperglycemia has independent risk factors in the development of macro-vascular complications of diabetes mellitus (DM) [13].

A sudden increase in blood glucose levels, which causes hyperglycaemia in type 2 diabetes patients, occurs due to the hydrolysis of starch by pancreatic α-amylase and the uptake of glucose by intestinal α-glucosidases [14]. An effective strategy for type 2 diabetes management is the strong inhibition of intestinal α-glucosidases and the mild inhibition of pancreatic α-amylase [15]. α-Amylase is present in both salivary and pancreatic secretion and is responsible for cleaving large malto-oligosaccharides to maltose, which is then a substrate for intestinal α-glucosidase. Tests for the ability of extracts and compounds to inhibit both α-amylase and α-glucosidase have been described by several workers.

As a number of anti-hyperglycemic agents have been found in plants, research into understanding the scientific basis for plant-based traditional medicines from various cultures has increased as scientists search for clues to discover new therapeutic drugs for type 2 DM [16-18]. Streptozotocin is an antimicrobial agent and has also been used as a chemotherapeutic alkylating agent and reported as diabetogenic. Again, this insulinopenia syndrome, called 'streptozotocin diabetes', is caused by the specific necrosis of the pancreatic beta cells and streptozotocin has been the agent of choice for the induction of diabetes mellitus in animals ever since [19].

In traditional Indian medicines plant and herbal extracts have been used as anti-diabetic agents [20]. These plants are typically rich in phenolic compounds, which are known to interact with proteins and can inhibit enzymatic action [21,22]. A number of medicinal plants and herbal extracts have been found to inhibit the enzymatic activity of α-glucosidase and α-amylase, and may have the potential as dietary anti-diabetic agents to control post-prandial hyperglycemia [23-26].

METHODOLOGY

Materials and Methods:

α-Amylase,α-glucosidase and streptozotocin were purchased from Sigma Aldrich India Pvt. Ltd. Laboratory grades reagents were used in all experiments. The instrument, UV spectrophotometer (UV 1700 Simazdu) was used in experiments.

Animals:

All the experiments were carried out using Wistar albino rats and were obtained from animal house, Birla Institute of Technology, Mesra, Ranchi, India. The animals were kept in polypropylene cages with paddy husk as bedding. Animals were housed at a temperature 24± 2°C and relative humidity of 60-70 %. A 10:14 light: dark cycle was followed. All animals were allowed to free access to water and fed with standard commercial rat chow pallets (M/s. Hindustan Lever Ltd, Mumbai). All the experimental procedures and protocols used in this study were reviewed by the Institutional Ethics Committee (Reg. No:621/02/ac/CPCSEA) and were in accordance with the guidelines of the CPCSEA, Ministry of Forests and Environment, Government of India.

Effect of Scopariadulcis on the activity of α-amylase

1g of the *S. dulcis* extract was mixed with 100 ml of the 2% potato starch solution in a 200 ml beaker. 1g of α-amylase was then added to this mixture and stirred vigorously at 37°C. After 30 min, 0.1 mol/l sodium hydroxide was added to terminate the α-amylase activity. The glucose content of the solution was determined by the above mentioned procedure [13].

Inhibitory assay of α-glucosidases enzyme

The inhibitory activity was determined by incubating a solution of starch (2% w/v maltose or sucrose) 1 ml with 0.2 M Tris buffer pH 8.0 and various concentration of plant extract for 5 min at 37^0C. The reaction was initiated by adding 1 ml of α-glucosidase enzyme (1U/ml) to it followed by incubation. Then, the reaction mixture was heated in boiling water bath to stop the reaction. The amount of glucose liberated is measured by glucose oxidase peroxidase method [27].

Calculation of 50% inhibitor concentration (IC_{50}):

The IC_{50} values (inhibitor concentration at which 50% inhibition of the enzyme activity occurs) of the plants extract were determined by performing the assay as above with varying concentration of the plant extracts ranging 20 to 100 µg/ml. The IC_{50} values were determined from plots of percent inhibition vs log inhibitor concentration and calculated by non-linear regression analysis from the mean inhibitory values [28].

Table 1. The percent α-amylase inhibition of MESD at varying concentrations

Plant	Concentration/ percentage of inhibition				
	20 (μg/ml)	40 (μg/ml)	60 (μg/ml)	80 (μg/ml)	100(μg/ml)
MESD	38.7	42.8	48.2	53.8	58.9
Acarbose	39.2	44.6	52.6	58.6	66.7

Table 2. The percentage inhibition of α-glucosidase of MESD at varying concentrations

Plant species	Concentration (μg/ml)	Inhibition (%)	IC_{50}
MESD	20	38.0	80.35 (μg/ml)
	40	42.1	
	60	46.4	
	80	50.8	
	100	54.3	

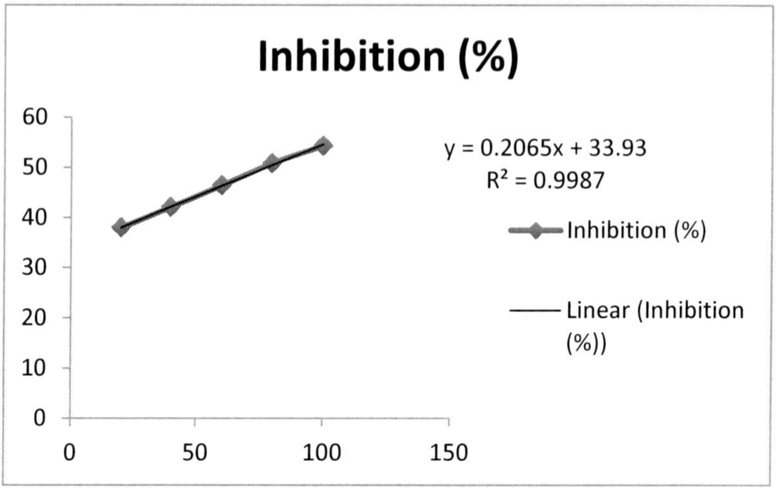

Figure1. IC_{50} value appear at 50% inhibition

Streptozotocin induced Anti-diabetic Activity of *Scoparia dulcis*

Determination of Acute Toxicity:

Acute oral toxicity was performed in mice by following Organization for Economic Cooperation and Development (OECD) guidelines [29,30].

Streptozotocin induced Anti-diabetic activity:

The whole study was divided in four animal groups. Group I received vehicle only (Tween-80, 1%) and served as control. Group II and group III received methanolic extract of *Scoparia dulcis* (MESD) 200 mg/kg body weight (BW) low dose and 400 mg/kg BW therapeutic dose and group IV *Glibenclamide* (1.0 mg/kg) BW serve as standard (p.o.)

respectively. 60 mg/kg BW streptozotocin was injected to every group animals following 18 h fast. After 48 h the blood glucose of surviving rats were determined by glucose oxidase method. Only those animals which showed hypoglycemia (blood glucose levels = 200-500 mg/dl), were considered diabetic and taken for further experimentation. The study was carried out after repeated administration of the extract for 21 consecutive days. Basal glycemia was determined in overnight fasted animals. The vehicle (1% Tween-80), MESD (200 and 400 mg/kg) and glibenclamide (1.0 mg/kg) were orally administered to the animals of group I-IV for 21 consecutive days. Blood samples collected on the 7th, 14th and 21st days were analysed for the determination of blood glucose [31].

Statistics:

The results were presented as mean ±SEM. "One-way Anova with Dunnett'spost test was performed using Graph Pad Prism version 3.00 for windows. Graph Pad Software, San Diego California USA, P<0.01 were considered significant.

In-vivo study of MESD was done on streptozotocin induced diabetic rats. The effect of the treatment with extract and Glibenclamide on blood glucose concentration in diabetic rats after post-treatment days was shown in Table 3. The study was conducted by blood collection on the 7th, 14th and 21th days after induces of diabetes and analyzed for the determination of blood glucose. All the values were found significant in both treatment dose (200 and 400mg/kg) as compared to control and standard Glibenclamide.

Table 3. Effect of repeated oral administration of extracts and (1.0mg/kg) glibenclamideon blood glucose and serum insulin level in streptozotocin diabetic rats

Treatment	Post-treatment (day) Blood glucose profile (mg/dl)			
	0	7th	14th	21st
Control (vehicle)	342.20±15.01	421.00±24.31	395.00±24.31	419.00±24.44
MESD200 mg/kg	331.00±21.46	316.80±13.60**	302.80±16.60**	279.40±13.55**
MESD 400 mg/kg	319.00±16.46	275.80±16.60**	259.80±16.60**	231.40±13.55**
Glibenclamide 1mg/ kg	311.00±9.44	251.00±17.84**	231.00±15.84**	219.00±11.79**

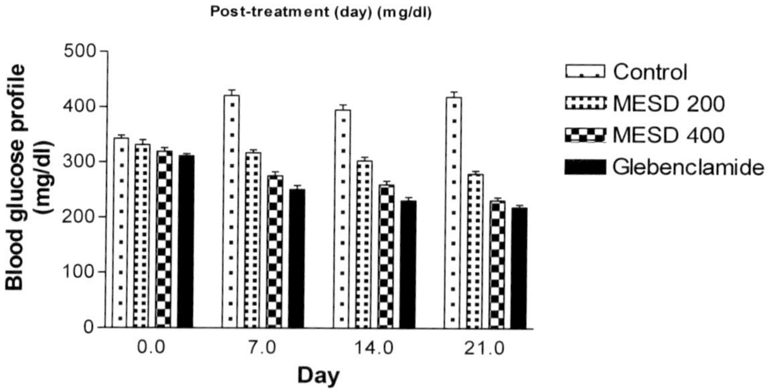

Figure 2. Effect of oral administration for 21 days of extracts and glibenclamide in streptozotocin diabetic rats

Antioxidant Activity

There are clear cut evidences to show the role of free radicals in diabetes and studies indicate that tissue injury in diabetes may be due to free radicals. The significance of oxidative stress in the disease pathology is uncertain but is frequently proposed to be related to the hyperglycemia. Other possible sources include elevated plasma lipids leading to increased lipid oxidation and decreased levels of antioxidant defense systems [32,33].

Aim of this study is to evaluate the antioxidant activity by using different models of MESD & WESD.

Materials and Methods:

Scopariadulcis whole plant was collected from waste field of Barpali, Odisha, India. 1, 1-Diphenyl-2, picrylhydrazyl (DPPH) was purchased from siga Aldrich India pvt Ltd. Laboratory grades reagents were used in all experiments. The instruments, UV spectrophotometer (UV 1700 Simazdu) and electrochemical analyzer (680B CH Instruments, USA) were used in experiments.

Extraction of Total Extractable Polyphenol (TEP)

The powdered sample (500 mg) was extracted sequentially with 40 ml methanol: water (50:50 v/v) and 40 ml acetone: water (70:30 v/v) at room temperature for 60 minutes each treatment. Supernatants from these extractions were combined and concentrated to a small volume under reduced pressure at 50°C, and made up to 10 ml with the same solvents (v/v). TEP were assayed spectrophotometrically by the Folin-Ciocalteau method using gallic acid as a standard and expressed as gallic acid equivalents [34]. The principle of reaction is that the phenols react with phosphomolybdic acid in Folin-Ciocalteau reagent in alkaline medium and produce blue coloured complex (molybdenum blue).

Methods of Antioxidant Activity

The antioxidant activity was done by three methods:

1. DPPH.
2. Reducing Power.
3. Electrochemical measurements.

DPPH

The antioxidant activity of the TEP, based on the scavenging activity of the stable 2, 2-diphenyl-1, picrylhydrazyl (DPPH) free radicals, was determined by the method described by Kumaran*et al* [35]. MESD & WESD TEP (0.1 ml) was added to 3 ml of a 0.004% MeOH solution of DPPH. Water (0.1 ml) in place of TEP was used as control. Absorbance at 517 nm was determined after 30 min. Graph was plotted between absorbance and concentration with TEP and gallic acid taken as standard anti-oxidative compound.

Reducing Power Determination

The reducing power of TEP of MESD & WESD was determined according to the method described by Chartarjee*et al* [36]. Different amounts of TEP (100-300 μl) were mixed with phosphate buffer (2.5ml, 0.02 M, pH 6.6) and potassium ferricyanide [K_6Fe (CN) $_6$] (2.5 ml, 1%). The mixture was incubated at 50°C for 20 min. A portion (2.5 ml) of trichloroacetic acid (10%) was added to the mixture, which was then centrifuged 3000 rpm for 10 min. The upper layer of solution (2.5 ml) was mixed with distilled water (2.5 ml) and $FeCl_3$ (0.5 ml, 0.1%). A higher absorbance indicates a higher reducing power.

Electrochemical Measurements

Electrochemical Measurements were conducted at 20 ± 0.1°C in a conventional three-electrode cell, cyclic voltammeter [PGSTAT model 30 (Ecochemie, Twente, Holland)] controlled by GPES 3.2 software from Ecochemie, running on a PIV personal computer]. In voltammetric test, the working electrode was a glassy carbon electrode; counter electrode was a platinum wire, while the reference electrode was an aqueous Ag/AgCl electrode used for our experiment. All tests potential was scanned from 0.1 to 1.0 V at a sweep rate of 100 mVs^{-}. Test solutions were prepared by MESD & WESD in 0.1 mol dm^{-3} KCl and made up the volume to 10 ml [37].

Absorbance vs concentration of methanol & water extract of *S.dulcis* & Gallic a

Concentration mg/ml	Absorbance		
	Gallic Acid	MESD	WESD
10	0.283±0.008	0.260±0.011	0.260±0.011
25	0.176±0.014	0.153±0.008	0.136±0.012
50	0.076±0.008	0.080±0.005	0.076±0.006
75	0.043±0.006	0.033±0.002	0.025±0.002
125	0.050±0.005	0.034±0.002	0.023±0.001
250	0.050±0.005	0.033±0.002	0.024±0.002

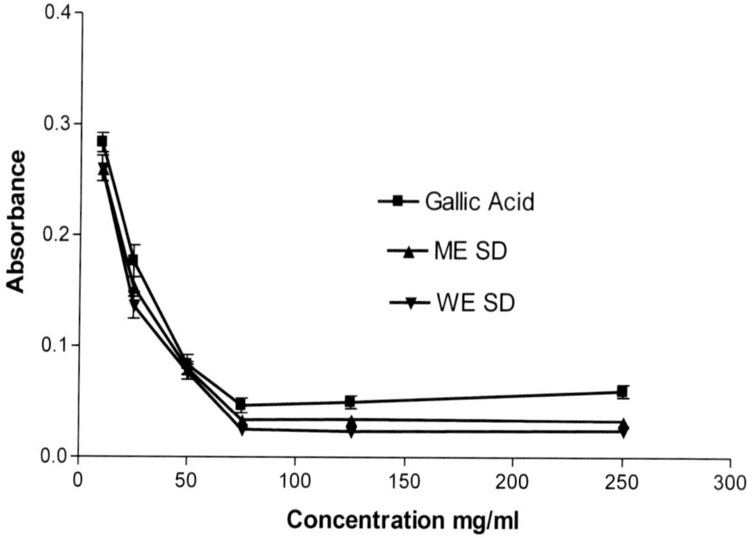

Figure 3. Plot of Absorbance verses concentration of methanolic& water extract of *S.dulcis* and Gallic acid.

Table 5. Reducing power of methanolic& water extract of *S.dulcis* and Gallic acid.

Concentration mg/ml	Absorbance		
	Gallic Acid	MESD	WESD
50	0.400±0.011	0.260±0.011	0.220±0.011
100	0.496±0.020	0.376±0.014	0.300±0.011
200	0.636±0.008	0.470±0.005	0.366±0.006
300	0.740±0.011	0.650±0.011	0.533±0.017

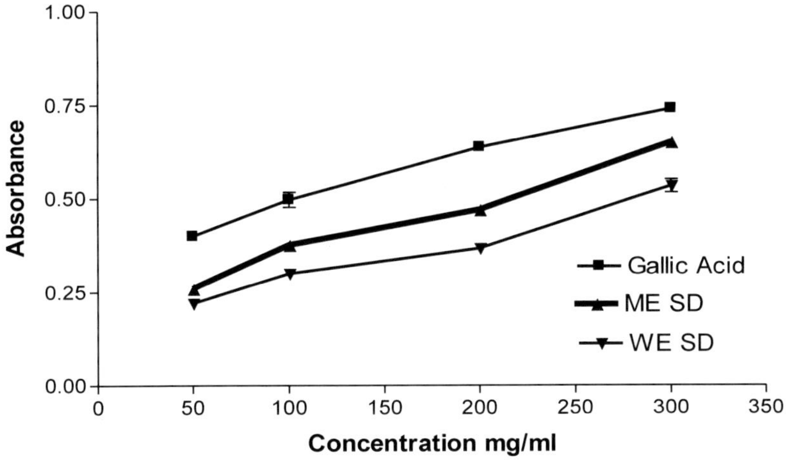

Figure 4. Plot of reducing power of methanolic& water extract of *S.dulcis* and Gallic acid.

Antihypertensive Activity

Cardiovascular disease (CVD) is a major health problem across the world, accounting for 30% of all deaths [39]. CVD has reached epidemic proportions in India [40] and is estimated to result in more than 3 million deaths each year [41]. Hypertension is a multifactorial disease dependent on complex interactions between genetic and environmental factors, yet many of these causes are not completely understood. Hypertension is a major risk factor for cardiovascular diseases and a large body of evidence suggests oxidative stress, an increase in the production of reactive oxygen species (ROS) as a strong underlying factor in hypertension [42-44].

This brief review will examine a common experimental model of hypertension, namely the effects of chronic administration of deoxycorticosterone acetate (DOCA), reduced renal mass and a high salt diet (DOCA–salt model) characterized by neurohumoral activation and volume expansion [45] . The DOCA–salt model is similar to other experimental models that depend on high salt intake and reduced renal mass leading to hypervolaemia. Excess DOCA, a mineralocorticoid leads to an imbalance in renal sodium handling by increasing sodium and water reabsorption in the distal tubules in dogs [46]. This is exacerbated by excess salt intake and reduced renal mass leading to an increase in extracellular fluid and plasma volume [47]. Deoxycorticosterone acetate–salt hypertension is thought to occur in several stages [48] described an early phase (days) of DOCA treatment where sodium is retained at the expense of potassium followed by a second phase (weeks) characterized by sodium balance and chronic potassium depletion. Acute extracellular volume expansion lowers the blood pressure of early phase hypertensive rats [49] suggesting dilution of elevated sodium levels. Titze*et al.* (2005) [50] reported that DOCA–salt rats had excess total body sodium with moderate water retention. There is an initial increase in fluid and salt intake that, for the most part, is matched by an increase in sodium and water excretion [51]. Recently, investigators have proposed that DOCA–salt hypertension in rats develops in stages with an abrupt increase in arterial pressure during the first 48 h followed by a delayed, slower rise in arterial pressure over the next few weeks leading to sustained hypertension [52,53]. Others have described a later phase, often termed malignant, where severe hypertension is observed 8–12 weeks later. In this review, we will discuss the evidence for and against the role of neural and humoral factors during early, developed and malignant DOCA–salt hypertension [54].

Angiotensin-converting enzyme (ACE) inhibitors are used to treat many conditions such as hypertension (high blood pressure) and some heart conditions. They block a protein in the body that leads to tightened blood vessels. As a result, blood vessels are more relaxed, lowering blood pressure. Several ACE inhibitors such as captopril, enalapril, lisinopril and temocapril are in clinical use for the treatment of hypertension. All of these drugs produce side effects, thus justifying the search for natural ACE inhibitors for safe andeconomical use. A number of compounds from plants have been identified to possess *in vitro* ACE inhibitor activity, including hydrolysable tannins, phenylpropanes, proanthocyanidins, flavonoids, xanthones, fatty acids, terpenoids, alkaloids, oligosaccharides and peptide amino acids, among others [55].

Hypertension is potentiated by angiotensin II, a potent vasoconstrictor agent [56]. Angiotensin II is formed from angiotensin I, a histidyl-leucine dipeptide, by the action of angiotensin I-converting enzyme (ACE) [57]. Anti-hypertensive drugs have been isolated from bovine beta-lactoglobulin and a number of plantsextract [58-61]. *Scoparia dulcis* is a rich source of isoflavonoidphenolics especially genistein and daidzein which have been shown to have numerous potential health benefits [62,63,73]. A number of flavonoids and isoflavonoids, including genistein have been shown to inhibit α-glucosidaseactivity *in vitro* [64].

Therefore, phenolic-rich extracts of *Scoparia dulcis* may have potential as a source of DOCA salts and anti-ACE agents for control of hypertension, a known complication of long-term diabetes and/or hyperglycemia. Here, methanolic extract phenolic-optimized extracts of *Scoparia dulcis* was assayed *in vivo* DOCA salts and *in vitro* and rabbit lung ACE.

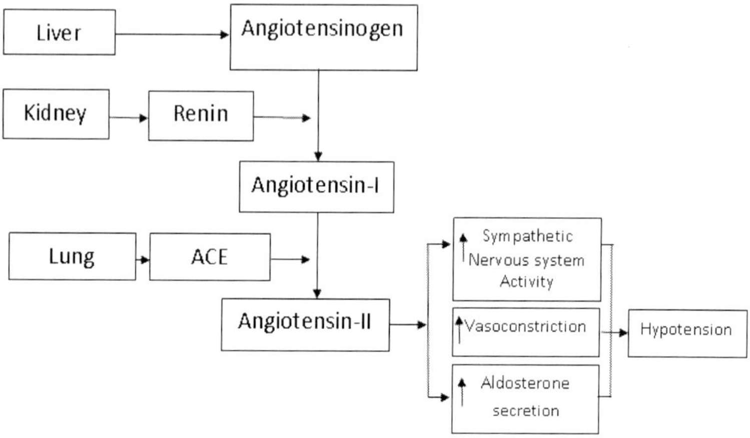

Figure 5. Flow chart of Mechanism of ACE inhibition

ANTIHYPERTENSIVE ACTIVITY in vitro

ACE-inhibitory activity

Materials and Methods:

Enzymatic Assay of ACE (EC 3.4.15.1)

ACE-inhibitory activity of hydrosylate sample was determined in triplicate by a method adopted from LO & Li-Chan (2005) [65] based on ACE activity assay of Cushman and Cheung (1973) [66]. In this assay, ACE activity is quantified by spectrophotometric absorbance at 228 nm produced by the liberation of hippuric acid (HA) from synthetic substrate hippuryl-L-histidyl-L-leucine (HHL). In the presence of ACE inhibitor at given concentration, ACE activity and HA production is depressed; therefore % ACE inhibition can be calculated as:

$$1-\left(\frac{(absorbanceofinhibitorcontainingsample - absorbanceofnegativecontrol)}{(absorbanceofpositivecontrol - absorbanceofnegativecontrol)}\right)X100$$

Where the positive control is taken as 100% ACE activity, with inhibitor having been replaced with buffer and the negative control is taken as 0 % ACE activity, with ACE being added only after reaction termination. The IC_{50}value, the concentration of peptide sample in the assay required to inhibit the activity by 50%, is commonly used to quantitatively express and compare potency of ACE-inhibitory activity between samples. In this study, IC_{50} values of methanol and water extract are shown in Figure 6. All controls were also performed in triplicate to monitor the reproducibility of the assay.

To carry out the assay, 0.01-0.25 mg/ml solutions of MESD and WESD were first mixed with 30 μL (2.5 mU) of ACE incubated at 37°C for 1 hour, after which 150 μL of 7.8 mM HHL was added and the solution was further incubated at 37°C for 1 hour. All assay components were dissolved in 0.025 M sodium tetraborate buffer containing 0.3 M NaCl (pH 8.3) for a total assay volume of 210 μL. HCl (250 μL, IN) was used to terminate the reaction. In order to extract HA, 1.0 ml of ethyl acetate was added. The solutions were then vortexed for 30 seconds (CM 101 Cyclo Mixer, Remi, India) and centrifused at 2000 x g (Remi-C24BL India) for 5 minutes. After centrifugation, 0.7 ml of ethyl acetate layer was removed into a 10 mm diameter clear glass tube and was evaporated by heating at 120°C in a hot air oven for 30 minutes. The remaining HA residue was redissolved with 1.3 ml of distilled water and the absorbance was read at 228 nm in a quartz cuvette (UV 1700 Shimazhu, Japan)

Where the positive control is taken as 100% ACE activity, with inhibitor having been replaced with buffer and the negative control is taken as 0 % ACE activity, with ACE being added only after reaction termination. The IC_{50} value, the concentration of peptide sample in the assay required to inhibit the activity by 50%, is commonly used to quantitatively express and compare potency of ACE-inhibitory activity between samples. In this study, IC_{50} values of methanol and water extract, as shown in Fig 6 & Table 6. All controls were also performed in triplicate to monitor the reproducibility of the assay.

Table 6. ACE activity of MESD & WESD and their IC_{50}

Dose mg/ml	%ACE activity of MESD	IC_{50}	%ACE activity of WESD	IC_{50}
0.01	95.66±0.88	0.102 mg/ml	97.66±0.88	0.124 mg/ml
0.05	66.33±1.85		77.33±1.20	
0.10	42.00±1.15		55.00±0.57	
0.15	23.33±0.88		32.33±0.88	
0.20	11.00±0.57		18.33±0.88	
0.25	7.66±0.33		13.66±0.88	

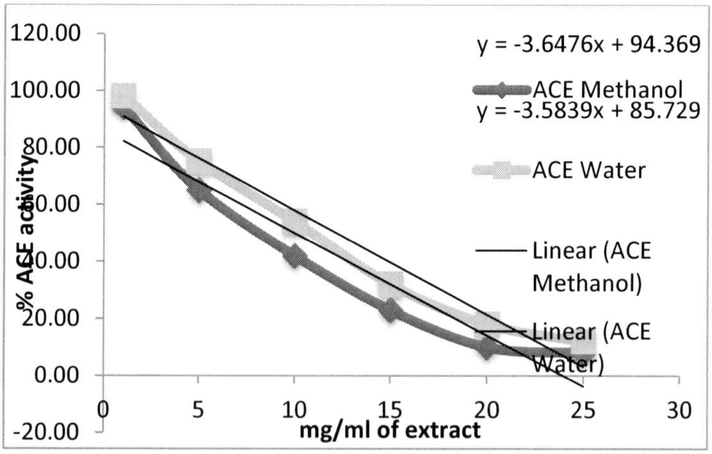

Figure 6. % ACE activity of MESD & WESD

ANTIHYPERTENSIVE ACTIVITY in vivo

Materials & Methods:

DOCA-salt induced hypertension:

Animals:

All the experiments were carried out using Wistar albino rats and were obtained from animal house, Birla Institute of Technology, Mesra, Ranchi, India and kept in polypropylene cages with paddy husk as bedding. Animals were housed at a temperature24± 2°C and relative humidity of 60-70 %. A 10:14 light: day cycle was followed. All animals were allowed to free access to water and fed with standard commercial rat chow pallets (M/s. Hindustan Lever Ltd, Mumbai). All the experimental procedures and protocols used in this study were reviewed by the Institutional Ethics Committee (Reg. No:621/02/ac/CPCSEA) and were in accordance with the guidelines of the CPCSEA, Ministry of Forests and Environment, Government of India.

Dosing of Animals:

The group I received vehicle only (Tween-80, 1%) and served as control. Group II and group III received 200 mg/kg BW low dose and 400mg/kg BW therapeutic dose (oral), respectively.

DOCA-salt induced hypertension and blood pressure measurements:

- Wistar rats were divided into four groups (n = 6). Normal control (NC) rats received a saline injection (0.5 ml/kg BW, twice weekly for 5 weeks, s.c., n = 6) and were age matched with treated rats. DOCA salt injection (20 mg/kg BW, twice weekly for 5 weeks, s.c., n = 6) was given to the rats and NaCl (1%, w/v) was added to their drinking water [67].

- DOCA salt-induced hypertensive rats were divided into 4 groups, viz. DOCA salt hypertensive control, test 1 and 2 compound treated and standard drug treated rats. DOCA salt hypertensive control rats received saline injection (for 12 days, n = 6).
- Test compounds treated rats received methanolic extract of Scopariadulcis (MESD) (200 mg and 400 mg/kg BW, for 12 days daily, i.p., n = 6) and standard drug treated rats received nifedipine (2 mg/kg BW, for 12 days daily, i.p., n = 6).
- Rats were initially trained for systolic blood pressure measurement on at least three separate occasions to establish a baseline blood pressure. Systolic pressure was measured by tail-cuff method using Non-invasive blood pressure instrument (NIBP-IICT Inc., Germany).

Table 7. Changes in mean systolic blood pressure after 12 days of Extract (200 and 400 mg/kg) treatment in DOCA-salt hypertensive rats

Treatment	**Mean Blood Pressure (Mean ± S.E.M.)**
Normotensive	115 ± 3.19
DOCA-Hypertensive	192 ± 4.95
MESD 200 mg/kg BW	156 ± 3.87**

MESD 400 mg/kg BW	139 ± 5.78**
Nifedipine treated 2mg/kg BW	121 ± 4.19**

Cumulative values are reported as mean±S.E.M. for six rats in each group. *$P < 0.05$; **$P < 0.01$ compared with DOCA-Hypertensive using; a one-way analysis of variance (ANOVA) followed by Dunnett's test.

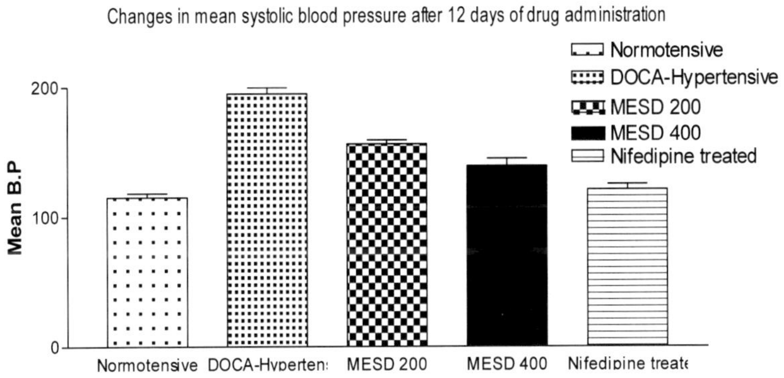

Figure 7. Changes in mean systolic blood pressure after 12 days of Extract (200 and 400 mg/kg) treatment in DOCA-salt hypertensive rats

DISCUSSION

In vivo study of MESD was done on streptozotocin induced diabetic rats. The effect of the treatment with extract and Glibenclamide on blood glucose concentration in diabetic rats after post-treatment days was shown in Table 3. The study was conducted on the 7th, 14th and 21th days after induction of diabetes and analyzed for the determination of blood glucose level. All the values were found significant in both treated doses (200 and 400mg/kg) as compared to control and standard.

The TEP of MESD & WESD was determined as gallic acid equivalent and the results were found to be 32.24 & 38.32 mg/g of gallic acid respectively.

The TEP of MESD & WESD showed a concentration dependent antiradical activity by inhibiting DPPH radical as equivalent to gallic acid. DPPH is usually used as a substrate to evaluate antioxidative activity of antioxidants. The method based on the reduction of methanolic DPPH solution in the presence of a hydrogen donating antioxidants due to the formation of the non radical form DPPH-H by the reaction DPPH to the yellow coloured diphenylpicrylhydrazine. It has been found that cystine, glutatathione, ascorbic acid, tocopherol, polyhydroxy aromatic compounds (e.g., hydroquinone, pyrogallol, gallic acid) and aromatic amines (e.g., *p*-phenylenediamine, *p*-aminophenol) reduced and decolorized 1, 1-Diphenyl-2, picrylhydrazyl by their hydrogen donating ability. It appears that the TEP of MESD & WESD possess hydrogen donating capabilities and act as an antioxidant. The scavenging effect increased with increasing activity of gallic acid, a known antioxidant used as positive control gave similar result as compared to TEP of MESD & WESD.

The defence mechanismof our body prevents us from the hazardous effects of free radicals by reducing these molecules with the antioxidant substances. The Fig.3& 4showed the reductive properties of sample extracts as compared to gallicacid,an established standard. The reductive ability was measured by Fe^{3+}-Fe^{2+} transformation in the presence of TEP of

MESD & WESD. Earlier we have observed a direct correlation between antioxidant activities and reducing power of certain plant extract [19]. The reducing properties are generally associated with the presence of reductones, which have been shown to exert antioxidant activity. The reductones break the free radical chain by donating a hydrogen atom [38]. The present data on the reducing power of TEP of MESD & WESD suggested that it was an active antioxidant like gallic acid.

The potential antihypertensive activity of *Scoparia dulcis* was evaluated by the inhibition of the angiotensin converting enzyme (ACE), using a UV Visible spectroscopy. The IC_{50} value of MESD was found to be 0.102mg/ml and WESD 0.124mg/ml (Table 6 and Figure 6&7). These values indicate that *Scopariadulcis* has a good ACE inhibitor activity.

The MESD showed higher IC_{50} in comparison to WESD and for this reason we have selected methanolic extract for its *in-vivo* study. It was observed that the *Scoparia dulcis* drug have the activity on the systolic blood pressure in case of DOCA induced hypertension in rats. All the value of tested MESD (200 and 400mg/kg) was found significant ($p<0.01$) systolic BP value in comparison to DOCA induced hypertension (Table 7 and figure-7).

The ACE inhibitor activity of *Scoparia dulcis* may be due to the presence of its chemical constituents like coumarins, phenols, saponins, tannins, amino acids, flavonoids, terpenoids and catecholamines [68]. Compounds belonging to these classes of natural products have been described to possess ACE in vitro inhibiting activity and, for this reason, *Scoparia dulcis* was selected for antihypertensive activity [69]. It contains terpenoids, which are responsible for numerous medicinal effects. Scoparic acid A, scoparic acid B, scopadulcic acid A and B, scopadulciol and scopadulin are all biologically active. Additional identified terpenoids of broomweed include alpha-amyrin, betulinic acid, dulcioic acid, friedelin, glutinol and ifflaionic acid [70].

The *Scoparia dulcis* showed presence of tannins, flavanoids and triterpine that have good antioxidant activity, might be due to the antioxidant ability of this compound shown to reduce hypertension in experimental animal models. Several cohort studies have suggested that high intake of flavonoids may decrease the risk of coronary heart diseases [71,72]. These findings suggest that systolic blood pressure has been remarkably controled by treatment of MESD in DOCA salt hypertensive rats.

CONCLUSION

The antidiabetic study both on *in vitro* and *in vivo* models and the antioxidant activity of MESD and WESD for comparison, showed positive result to provide substantiation for its antidiabetic activity, because there is a link between the antioxidant and the antidiabetic activity. The antihypertensive activity for both *in vitro* and *in vivo* models was carried out. *In in vitro* models on MESD and WESD with MESD showing significant reduction in systolic blood pressure on DOCA induced hypertensive rats.

REFERENCES

[1] Mukherjee PK, Wahile A. Integrated approaches towards drug development from Ayurveda and other Indian system of medicines. Journal of Ethnopharmacology 2006;103:25-35.

[2] World Health Organisation. NewDelhi: Regional office for South East Asia; 2000. p. 3.

[3] Mukherjee PK. Exploring botanicals in Indian system of medicine-Regulatory perspectives. Clinical research and regulatory affairs 2003;20:249–264.

[4] Blumenthal M. A National Curriculum for Medical Students, American Medical Student Association (AMSA) Foundation 3 EDCAM Project: AmsaEdcam Initiative; 2002. p. 3-6.

[5] Khare CP. Indian medicinal plants: An illustrated dictionary. Heidelberg: Springer-Verlag; 2007. p. 592.

[6] Subbarayappa BV. Siddha medicine: an overview. Lancet 1997;350:1841-1844.

[7] World Health Organization, WHO Traditional Medicine Strategy 2002- 2005, WHO, Geneva, Switzerland, 2002.

[8] Loizzo MR, Antoine MS, Tundis R, Menichini F, Bonesi M, Piccolo V, Giancarlo A, de Cindio SB, Houghton PJ, Francesco M. In vitro inhibitory activities of plants used in Lebanon traditional medicine against angiotensin converting enzyme (ACE) and digestive enzymes related to diabetes. Journal of Ethnopharmacology 2008;119:109-116.

[9] Harrower AD. Comparison of efficacy, secondary failure rate, and complications of sulfonylureas. J Diabetes Complications 1994; 8:201-203.

[10] Reuser AJ, Wisselaar HA. An evaluation of the potential side-effects of alphaglucosidase inhibitors used for the management of diabetes mellitus, Eur J Clin Invest 1994; 24:19-24.

[11] Campbell RK, White JR, Saulie BA. Metformin: a new oral biguanide. ClinTher1996; 18:360-371.

[12] Kwon YI, Apostolidis E, Shetty K. Inhibitory potential of wine and tea against α-amylase and α-glucosidase for management of hyperglycemia linked to type 2 diabetes.J Food Biochem2008; 32:15–31.

[13] Mishra A, Jha S. In vitro Postprandial Glucose lowering effects of Dietary fibers Isolated from Tamarindusindica and Cassia fistula seeds. American J. of Food Tech 2011;6:435-440.

[14] Gray G. Carbohydrate digestion and absorption–role of small intestine. N Engl J Med 1975; 292:1225-1230.

[15] Krentz AJ, Baile CJ. Oral antidiabetic agents: Current role in type 2 diabetes mellitus. Drugs 2005; 65:385-411.

[16] Jarvill-Taylor KJ, Anderson RA, Graves DJ. A hydroxychalcone derived from cinnamon functions as amimetic for insulin in 3T3-L1 adipocytes, J Am CollNutr 2001; 20:327-336.

[17] Maroo J, Vasu VT, Aalinkeel R, Gupta S. Glucose lowering effect of aqueous extract of EnicostemmalittoraleBlume in diabetes: a possible

mechanism of action, J Ethnopharmacol 2002;81:317-320.

[18] Elder C. Ayurveda for diabetes mellitus: A review of the biomedical literature. AlternTher Health Med 2003;10:44-50.

[19] S. Lenzen. The mechanisms of alloxan and streptozotocin induced diabetes. Diabetologia 2008;51:216-226.

[20] Grover JK, Yadav S, Vats V. Medicinal plants of India with anti-diabetic potential. J Ethnopharm 2002;81:81-100.

[21] Dawra RK, Makkar HP, Singh B. Protein-binding capacity of microquantities of tannins. Anal Biochem 1988;170:50-53.

[22] Suryanarayana P, Kumar PA, Saraswat M, Petrash JM, Reddy GB. Inhibition of aldose reductase by tannoidprinciples of *Emblicaofficinalis*: implications for the prevention of sugar cataract. Mol Vision 2004; 10: 148-154.

[23] Hanefeld M, Temelkova-Kurktschiev T. Control of post-prandial hyperglycemia – an essential part of good diabetes treatment and prevention of cardiovascular complications. NutrMetabCardiovasc Dis 2002; 12: 98-107.

[24] Vats V, Grover JK, Rathi SS. Evaluation of anti-hyperglycemic and hypoglycemic effect of *Trigonellafoenum-graecum* Linn, *Ocimum sanctum* Linn, and *Pterocarpusmarsupium* Linn in normal and

alloxanized diabetic rats, J Ethnopharm 2002; 79: 95-100.

[25] McCue P, Shetty K. Inhibitory effects of rosmarinic acid extracts on porcine pancreatic amylase in vitro. Asia Pac J ClinNutr 2004; 13 (1): 101-106.

[26] McCue P, Vattem D, Shetty K. Inhibitory effect of clonal oregano extracts against porcine pancreatic amylase in vitro, Asia Pac J ClinNutr 2004; 13 (4): 401-408.

[27] Andrade-Cetto A, Becerra-Jimenez J, Cardenas-Vazquez R. α-glucosidase inhibitory activity of some Maxican plants used in the treatment of type 2 diabetes. Journal of Ethno pharmacology 2008; 116: 27-32.

[28] Tadera K, Minaki Y, Takamatsu K, Matsuoka T. Inhibition of alpha glucosidase and alpha amylase by flavonoids, J NutrSciVitaminol 2006; 52: 149-153.

[29] Ecobichon DJ. The basis of toxicology testing, CRC Press, New York, 1997.

[30] Dixon RL, Hall JL. Reproductive Toxicology, In: Hayes AW, editor. Principles and Methods of toxicology, New York: Raven Press. 1982: 107-40.

[31] Bergymeyer HU, Benot E. Colorimetric methods with glucose oxidase and peroxidase. In: Kunst A, Draeger B, Ziegenhorn J, Bergymeyer H, editor. Methods of enzymatic Analysis, VerlagChemie GBH WellhheimBerostr, 3rd ed. Academic Press: New York; 1963. p. 179-185.

[32] Hunt JV, Smith CCT, Wolff SP. Auto-oxidative glycosylation and possible involvement of peroxides and free radicals in LDL modification by glucose. Diabetes 1990;39:1420-1424.

[33] Baynes P, Thorpe SR. Role of oxidative stress in diabetic complications: a new perspective on an old paradig. Diabetes 1999;48:1-9.

[34] Ruperez P, Saura-calixto F. Dietary fibre and physicochemical properties of edible spanish seaweeds. Euro food research Technol 2001;2K:344-354.

[35] Kumaran A, Karunakaran RJ. Antioxident and free redical scavenging activity of an aqueous extract of *Coleus aromaticus*. Food chemistry 2006;97:109-114.

[36] Chatterjee S, Niaz Z, Gautam S, Adhikari S, Variyar PS, Sharma A. Antioxidant activity of some phenolic constituents from green pepper (*Pippernigrum*L.) and fresh nutmeg mace (*Myristicafragrans*). Food Chemistry 2007;101:515-523.

[37] Gorden MH. The mechanism of antioxidant action in vitro. In Hudson BJF (Ed.), Food Antioxidants, London: Elseveir Applied Science 1990: 1-18.

[38] Gaziano TA. Economic burden and the cost-effectiveness of treatment of cardiovascular disease in Africa. Heart 2008; 94: 140-144.

[39] Gupta R, Joshi PP, Mohan V, Reddy KS, Yusuf S. Epidemiology and causation of coronary heart disease and stroke in India. Heart 2008; 94: 16-26.

[40] Prahalathan P, Kumar S, Raja B. Effect of morin, a flavonoid against DOCA-salt hypertensive rats: a dose dependent study. Asian Pacific Journal of Tropical Biomedicine 2012;443-448.

[41] World Health Organization. Mortality and burden of disease estimates for WHO member states in 2004. Geneva: WHO; 2004. Available from: http://www.whoint/entity/healthinfo/global_burden/disease/gbddeathdalycountryestimates 2004.xls. [Accessed on 20 July, 2009]

[42] Ceriello A. Possible role of oxidative stress in the pathogenesis of hypertension. Diabetes Care 2008; 31: 181-184.

[43] Cutler JA, Sorlie PD, Wolz M, Thom T, Fields LE, Roccella EJ. Trends in hypertension prevalence, awareness, treatment, andcontrol rates in United States adults between 1988-1994 and 1999-2004. Hypertension 2008; 52: 818-827.

[44] Schenk J, McNeill JH. The pathogenesis of DOCA-salt hypertension. J PharmacolToxicol Methods1992;27:161–170.

[45] Guyton AC, Coleman TG, Cowley AW, Scheel KW, Manning RD & Norman RA. Arterial pressure regulation: overriding dominance of the kidneys in long-term regulation and in hypertension. Am J Med 1972; 52: 584–594.

[46] Haack D, M¨ohring J, M¨ohring B, Petri M, Hackenthal E. Comparative study on development of corticosterone

and DOCA hypertension in rats. Am J Physiol Renal Physiol 1977; 233: F403–F411.

[47] M¨ohring J, M¨ohring B. Reevaluation of DOCA escape phenomenon. Am J Physiol1972; 223: 1237–1245.

[48] Villamil MF, Amorena C, Ponce J, Muller A, Taquini AC. Acute saline expansion lowers blood pressure of early DOCA-salt hypertensive rats. Pflugers Arch 1981; 390: 260–261.

[49] Titze J, Bauer K, Schafflhuber M, Dietsch P, Lang R, Schwind KH, Luft FC, Eckardt K, Hilgers KF. Internal sodium balance in DOCA-salt rats: a body compositionstudy. Am J Physiol Renal Physiol 2005; 289: F793–F802.

[50] Abrams JM, Osborn JW. Osmoregulatory function in health disease: a role for benzamil-sensitive proteins of the central nervous system in the pathogenesis of salt-dependent hypertension. ClinExpPharmacolPhysiol 2008;35:687–694.

[51] Brooks VL, Freeman KL, Qi Y. Time course of synergistic interaction between DOCA and salt on blood pressure: roles of vasopressin and hepatic osmoreceptors. AmJPhysiolRegulIntegr Comp Physiol 2006; 291: R1825–R1834.

[52] Jacob F, Clark LA, Guzman PA, Osborn JW. Role of renal nerves in development of hypertension in DOCA-salt model in rats: a telemetric approach. Am J Physiol Heart CircPhysiol 2005;289: H1519–H1529.

[53] Braga FC, Serra CP, VianaJúnior NS, Oliveira AB,Côrtes SF, LombardiJA. Angiotensin-converting

enzyme inhibition by Brazilian plants.Fitoterapia 2007; 78: 353–358.

[54] Probhakar YS, Gupta SP. Structure-activity relationship study on angiotensin-converting enzyme inhibitors - investigation of hydrophobic interaction in inhibition mechanism. Indian J BiochemBiophys 1985; 22: 318-320.

[55] Lieberman J. Elevation of serum angiotensin-convertingenzyme (ACE) level in sarcoidosis. Am J Med 1975; 59:365-372.

[56] Vermeirssen V, Van Camp J, Verstraete W. Optimisation and validation of an angiotensin-converting enzymeinhibition assay for the screening of bioactive peptides. J BiochemBiophys Methods 2002; 51: 75-87.

[57] Jonadet M. Bastide J, Bastide P, Boyer B, Carnat AP, Lamaison JL. Activitésinhibitricesenzymatiques in vitroetangioprotectrice in vivo d'extraits de karkadé (Hibiscus Sabdariffa L.). J Pharm Belg 1990; 45 (2): 120-124.

[58] Actis-Goretta L, Ottaviani JI, Keen CL, Fraga CG. Inhibition of angiotenisn converting enzyme (ACE) activity by flavan-3-ols and procyanidins. FEBS Lett 2003; 555:597-600.

[59] Kang DG, Lee YS, Kim HJ, Lee YM, Lee HS. Angiotensin converting enzyme inhibitory phenylpropanoid glycosides from Clerodendrontrichotomum. J Ethnopharm. 2003; 89:151-154.

[60] Hussain M, Banerjee M, Sarkar FH, Djuric Z, Pollak MN, Doerge D, Fontana J, Chinni S, Davis J, Forman J, Wood DP, Kucuk O. Soy isoflavones in the treatment of prostate cancer. Nutr Cancer 2003; 47 (2): 111-117.

[61] Magee PJ, Rowland IR. Phyto-oestrogens, their mechanism of action: current evidence for a role in breast and prostatecancer. Br J Nutr 2004; 91 (4): 513-531.

[62] Kim JS, Kwon CS, Son KH. Inhibition of alpha glucosidase and amylase by luteolin, a flavonoid. BiosciBiotechnolBiochem 2000; 64 (11): 2458-2461.

[63] Nishioka T, Kawabata J, Aoyama Y. Baicalein, an alphaglucosidase inhibitor from Scutellariabaicalensis. J NatProd 1998; 61 (11): 1413-1415.

[64] Lo WMY, Li-Chan ECY. Angiotensin I converting enzyme inhibitory peptides from in vitro pepsin-pancreatin digestion of soy protein. J Agric Food Chem 2005; 53: 3369-3376.

[65] Cheung HS, Cushman DW. Inhibition of homogeneous angiotensin converting enzyme of rabbit lung by synthetic venom peptides of Bothropsjararaca. BiochimBiophysActa 1973; 293: 451-463.

[66] Bankar GR, Nampurath GK, Nayak PG, Bhattacharya S. A possible correlation between the correction of endothelial dysfunction and normalization of high blood pressure levels by 1,3,4-oxadizole derivative, an L-type Ca2+ channel blocker in deoxycorticosterone acetate and NG-nitro-1-arginine hypertensive rats. Chemico-Biological Interactions 2010;183:327-331.

[67] Lee DS, Lee SH. Genistein, a soy isoflavone, is a potent alpha-glucosidase inhibitor. FEBS Lett 2001; 501 (1): 84-86.

[68] Gao H, Nishioka T, Kawabata J, Kasai T. Structure-activity relationships for alpha-glucosidase inhibition of baicalein, 5,6,7-trihydroxyflavone: the effect of A-ring substitution. Bio SciBiotechnolBiochem 2004; 68 (2): 369-375.

[69] Patrick McCue, Young-In Kwon and KalidasShetty. Anti-diabetic and anti-hypertensive potential of sprouted and solid-state bioprocessed soybean. Asia Pac J ClinNutr 2005;14 (2):145-152.

[70] Jalili T, Carlstrom J, Kim S. Quercetin-supplemented diets lower blood pressure and attenuate cardiac hypertrophy in rats with aortic constriction. J CardiovascPharmacol 2006; 47: 531-541.

[71] Mink PJ, Scrafford CG, Barraj LM, Harnack L, Hong CP, Nettleton JA, et al. Flavonoid intake and cardiovascular disease mortality: a prospective study in postmenopausal women. Am J ClinNutr 2007; 85: 895-909.

[72] Mishra MR, Behera RK, Jha S, Panda AK, Mishra A, Pradhan DK, Choudary PR. A Brief Review on phytoconstituents and Ethnopharmacology of Scopariadulcis Linn. (Scrophulariaceae). Int J PhytoMed 2011; 3 (4): 422-438.

CHAPTER THREE

THE INTERPLAY OF HIDDEN ORDER WITH SUPERCONDUCTIVITY IN HIGH-T_C SUPERCONDUCTORS

Kamal Lochan Mohanta *

Department of Physics, Siksha 'O' Anusandhan University, Bhubaneswar-751030, Odisha, India.

ABSTRACT

The mystery of pairing mechanism in high-T_c superconductors is still hidden. Here, we propose a microscopic model calculation of the interplay of superconductivity (SC) and spin density wave (SDW) in high-T_c systems. Moreover the charge density wave (CDW) is included in the Hamiltonian as a pseudogap interaction. The pseudogap co-exists in the superconducting state and extends to normal phase above T_c. These three temperature dependent order parameters indicating to SC, SDW and CDW are calculated by the Green's function technique by using the model Hamiltonian and are solved self-consistently. Finally, the conduction electron density of states(DOS) are calculated to interpret the scanning tunneling microscopy experimental data to explain the tunneling conductance of the system. The dip-hump structure as observed by the experiment is explained by our model calculation.

* Corresponding Author Email address:kamalmohanta@soauniversity.ac.in

Keywords: Superconductivity; Charge density wave; Spin density wave, Tunneling conductance

INTRODUCTION

Superconductivity: An overview

The quest for low temperature physics in the 19th century has shown the path for the discovery of superconductivity. Kamerlingh Onnes has discovered superconductivity in mercury at 4.2 K in his Leiden laboratory in 1911 [1]. The superconducting state is a state of matter below certain temperature where the dc resistivity suddenly drops to zero. This temperature is known as superconducting transition temperature (T_c). The materials, which display this phenomenon, are called superconductors. The superconductors display quantum behaviour at macroscopic scale.

In 1956 L. N. Cooper has showed that, in the presence of a very weak electron-phonon (lattice) interaction, two conducting electrons are capable of forming a stable paired state. This paired state is now referred to as the Cooper pair. The first microscopic theory of superconductivity in metals has been formulated by J. Bardeen, L. N. Cooper and R. Schrieffer in 1957, which is now known as the BCS theory [2]. The central concept of the BCS theory is a weak electron-phonon interaction which leads to the appearance of an attractive potential between two electrons. As a consequence, they form the Cooper pairs. This theory put a limit on the superconducting transition temperature not to surpass 30K. Based on this theoretical limit; superconductors are broadly classified into two categories after the discovery of copper oxide superconductors: (i) low T_c

superconductors (T_c < 25K) (ii) high T_c superconductors (T_c > 25K).

Many new superconductors have been discovered in 1970s and 1980s. Superconductivity in Nb_3Ge alloy was observed in 1973 with critical temperature 23.2 K which was the highest T_c achieved before the discovery of copper oxide superconductors [3]. Subsequently few new classes of superconductors like heavy fermions [4-6], organic superconductors [7] and carbon C_{60} compounds [8, 9] are also discovered.

The high-T_c superconductivity era has started in 1986, when Bednorz and Müller have found the evidence for superconductivity at T $\simeq$30K in La-Ba-Cu-O ceramics [10] which has crossed the limit of BCS theory. Shortly after this discovery, the groups at the Universities of Alabama and Houston under the direction of M. K. Wu and P. W. Chu have announced the discovery of a new rare earth based copper oxide superconductor Y-Ba-Cu-O [11] in 1987, which is the first material having a transition temperature (T_c $\simeq$93K) higher than the boiling point of liquid nitrogen. In recent years the discovery of the iron-arsenide layered superconductor, with high superconducting transition temperature, $T_c \simeq 56$K, has paved the new way in the experimental and theoretical research of superconductivity [12-14].

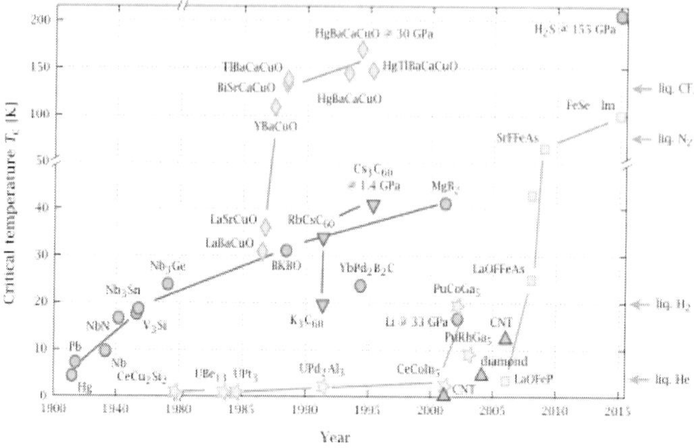

Fig. 1.1 Graphical presentation of T_c variation from its discovery till date.

As a summary, the time evolution of the superconducting critical temperature of various elements, alloys and compounds since the discovery of superconductivity in 1911 till date, is shown graphically in figure 1.1.

1.2 Discovery and Development of High-Tc Superconductors

J. G. Bednorz and K. A. Müller discovered the superconductivity in a ceramic copper oxide material at a temperature of about 30K [10] in 1986.The confirmation of these experiments [15] given a frantic race for the preparation of materials with even higher critical temperatures. A number of high-T_c compounds have been discovered in the recent years. The highest critical temperature for mercury-based material is about 137K [16]. On the experimental side results are being consolidated, mainly due to a considerable improvement in the quality of the samples compared with those used in the early studies of the high-T_c materials. Much work has been devoted to some unexpected properties of the cuprates, such as the linear behavior of the temperature dependent d.c. resistivity and a Hall coefficient, the presence of short coherence lengths, and the energy dependence of the relaxation rate $1/\tau$. The search for new materials has been mainly empirical, since no predictive theory is currently known for the high-T_c compounds. The fact is that no one knows why the cuprates behave as they do. High- T_c superconductors, such as $La_{2-x}Sr_xCuO_x$ (T_c $\simeq$K) and $YBa_2Cu_3O_{7-y}$ (T_c $\simeq$93K) were discovered in 1986 and 1987 respectively and have been actively studied since. The scientific community was amazed by the discovery of high-T_c superconductivity in copper oxides, because oxides are very bad conductors and phonon mediated superconductivity is impossible at so high temperatures. In early 1988, Bi- and Tl- based superconducting cuprates were discovered with T_cs equal to 110 and 125 K, respectively [17, 18]. Finally, Hg-based cuprates with the highest critical temperature T_c $\simeq$137K were discovered in 1993 and applying high pressure the T_c increased up to 164 K [19, 20]. These cuprates are hole-doped. One group of electron-doped cuprates: (Nd, Pr, Sm)-Ce-Cu-O was discovered in 1989. Their maximum critical temperature is comparatively low, $T_{c,max}$ $\simeq$24K [21, 22]. High-T_c physics has strongly attracted theorists to work on correlated electrons; but it may be claimed that, from a theorist's angle, the particular model or material is not as important as the development of the tools to handle them. Computational results can contribute to the acceptance or rejection of mean-field based theories, and can also indicate directions in which new approaches should be developed. Infact, the cross-fertilization between computational and analytical work in the area of correlated electrons is increasing rapidly.

1.3 Structure of High-T_c materials

The high-T_c materials are generally tetragonal, and all of them have one more CuO_2 planes in their structure, which are separated by layers of other atoms (Ba,O, La,.....). The presence of antiferromagnetic (AFM) order at low temperature in the undoped resime in high-T_c materials is another common property. The long range spin order is destroyed on doping and the superconducting phase appears. This does not mean that spin correlations are not important for superconductivity. The spin correlation length can be long in the superconducting (SC) phase even without strict long range order which producing a local arrangement of magnetic moments that at short distances differs very little from that observed below the N′eel temperature in the insulating regimes. In reminder of the section, we discuss the structure of some particular high-T_c compounds. The pure high-T_c materials such as the 214 compounds La_2CuO_4, Nd_2CuO_4 and $YBa_2Cu_3O_6$ are magnetic insulators. It is well known that undoped Nd_2CuO_4 is an antiferromagnetic insulator with a N′eel temperature $T_N \simeq 270K$. When doped with Ce, additional electrons enter into CuO_2 plane and the material becomes metallic. It is a superconductor with highest $T_c \simeq 20K$ at $x = 0.15$ [23]. Similarly the undoped La_2CuO_4 is a three dimensional antiferromagnetic insulator ($T_N = 300K$). When La is doped with Sr, it introduces holes in CuO_2 plane. The antiferromagnetic long range order disappears with increase in hole concentration and as a result N′eel temperature reduces drastically. There appears to be a spin glass state and then superconductivity sets in with $T_c = 37K$.

The compound $Bi_{2.2}Sr_2Ca_{0.8}Cu_2O_{8+\delta}$ with $T_c \simeq 84K$, has been found to be orthorhombic, with lattice parameters a = 5.414, b = 25.79, c = 30.89 A°, and with a substructure based on a unit cell of parameters a = 5.414, b = 5.418, c = 30.89A°. The structure of the atoms in the sub-cell shows layers of copper and oxygen perpendicular to the c-axis and quite similar to those observed in $YBa_2Cu_3O_7$. These layers are separated by others containing calcium, strontium and oxygen, and bismuth and oxygen. The bismuth is surrounded by six oxygen atoms in octahedral coordination, while the copper has a pyramidal environment similarly to that present in $YBa_2Cu_3O_7$.

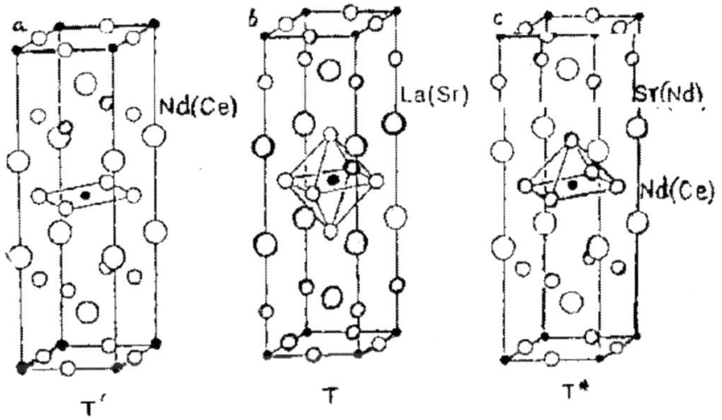

Fig. 1.2 The Crystal structure of (a) $Nd_{2-x}Ce_xCuO_4$ (T′-phases)

(b) $La_{2-x}Sr_xCuO_4$ (T-phases) and (c) $Nd_{2-x-x}Ce_xCuO_4$ (T^*-phases).

At first, we attempt to study the doped cuprates of the form $R_{2-x}M_xCuO_4$ (R=La, Nd and M=Ca, Ba, Sr, Ce). The lattice structure of three representatives, so called undoped parent compounds of the high-T_c superconductors (HTSC) are shown in the Fig.1.2(a,b,c). In essence, all the HTSC consist of two dimensional CuO_2 planes, which are sandwiched between the intervening atomic layers. All the cuprate superconductors have two dimensional character. This low dimensionality is confirmed from the following experimental observations:

(1) The band structure calculations indicate that the bands show very little dispersion along the crystal axis. This implies the two dimensional nature of electron states. (2)The structure measurements show that the distance between copper and oxygen atoms at the apexes of the Cu-O octahedron is large compared to that in the CuO_2 plane. These observations confirm their two dimensional character. It is experimentally well established that the parent compounds La_2CuO_4, Nd_2CuO_4 and $YBa_2Cu_3O_{6.5}$ [24-26] are insulators and ordered anti-ferromagnetically (AFM) below the N´eel temperature T_N. The superconductors are Ce4+ doped compounds, with the formula $R_{2-x}Ce_x$ CuO_{4-y}, where R stands for rare-earth i.e. La, Pr,Nd or Sm. The compounds have the Nd_2 $CuO_4(T')$ structure, which is composed of sheets of Cu − O squares (Fig.1.2.a); this structure has no apical oxygen atoms, in contrast to the T phase structure with Cu−O octahedral (Fig.1.2.b.), as observed in La_{2-x} Sr_x CuO_4 [27] and the T*-phase structure with Cu − O pyramids (Fig.1.2.c), as in Nd_{2-x-x} Ce_x Sr_xCuO_4 [27]. Electron doped superconductivity is not restricted to Nd-based compounds, but is also evident in the T′-phase of Pr- and Sm-based sheets of Cu-O squares (Fig.1.2.a); this structure has no apical oxygen atoms, in contrast to the T phase structure with Cu-O octahedral (Fig.1.2.b.), as observed in La_{2-x} Sr_x CuO_4 [27] and the T*-phase structure with Cu-O pyramids (Fig.1.2.c), as in $Nd_{2-x-x}Ce_xSr_xCuO_4$ [27]. Electron doped superconductivity is not restricted to Nd-based compounds, but is also evident in the T′-phase of Pr- and Sm-based copper oxide compounds with Ce doping.

1.4 Structural Instability in High- T_c Oxides

The presence of the structural instability inhibits even higher T_cs because of the destruction of partial Fermi surface (FS). The sudden change of the unit cell volume as a function of the charge transfer parameter in $HgBa_2CuO_{4+y}$ arrests the T_c growth [28]. Thermal expansion measurements on insulating La_2CuO_{4+y} and $La_{2-x}M_xCuO_4$ (M = Ba, Sr) with non-optimal doping show two lattice instabilities having $T_{d1} \simeq 32$ K and $T_{d2} \simeq 36$ K, while the under-doped Y $Ba_2Cu_3O_{7-y}$ with y =0.5 and T_c = 49 K has a single instability at $T_d \simeq 90$ K [29]. Both T_{d2} and T_d are close to maximal T_cs in the corresponding optimally doped compounds. Anomalies of the lattice properties above T_c in $La_{2-x}M_xCuO_4$ were also observed in ultrasound experiments (x = 0.14, M = Sr) as well as in thermal expansion and infrared absorption measurements. Resistive measurements of $La_{2-x}Sr_xCuO_4$ demonstrated an anomalous peak above T_c for superconducting samples with x = 0.06 and 0.075, persisting also for the semiconducting composition with x = 0.052, where the resistive upturn appears just below this peak . These authors consider the anomalies as a clear indication both of the structural and the electronic phase transitions to the more ordered charge stripe phase. Such anomalies in the vicinity of Tc were shown to be a rule for $La_{2-x}Sr_xCuO_4$, Y $Ba_2Cu_3O_{7-y}$ and Bi-Sr –Ca- Cu- O and cannot be explained by the superconducting transition per se. It should be noted that in addition to the doping independent transitions [29] in $La_{2-x}Ba_xCuO_4$ there are also successive transitions from high-temperature tetragonal (HTT) to low-temperature orthorhombic (LTO) and then to low-temperature tetragonal (LTT) phase with Tc suppressed to zero in the intermediate doping regime centered at x = 1, the superconducting region becoming doubly connected. At the same time, the $La_{2-x}Sr_xCuO_4$ phase diagram does not include a

LTT phase and its superconducting region is not broken by the normal state intrusion. $La_{2-x}Sr_xCuO_4$ doped with Nd leads to the LTT phase, and this kind of doping is widely claimed to provoke phase separation with either static or dynamic charged and magnetic stripes [30].

Returning to the $La_{2-x}M_xCuO_4$ family, it is important to point out that the atomic pair distribution functions in real space, measured by neutron diffraction both for M = Ba and Sr, revealed local octahedral tilts surviving even at high temperatures deep into the HTT phase. For $La_{2-x}Sr_xCuO_4$ with x = 0.115, electron diffraction disclosed that a low temperature structural transition is accompanied by the CDWs of the (1/2, 1/2, 0) type that lead to the suppression of superconductivity [31]. Raman scattering investigations indicated that in the under-doped case there is a pseudo gap $E_{ps} \simeq 700\ cm^{-1}$ without any definite onset temperature, whereas for the over-doped samples the pseudo gap is completely absent. On the other hand, EXAFS measurements for $La_{2-x}Sr_xCuO_4$ with x = 0.15 and La_2CuO_{4+y} with y = 0.1 demonstrated that CDWs and superconductivity coexist. In $Bi_2Sr_2CaCu_2O_{8+y}$, lattice scattering anomalies above Tc were observed in the same manner as in $La_{2-x}Sr_xCuO_4$ and $YBa_2Cu_3O_{7-y}$. It is remarkable that in $Bi_2Sr_2CaCu_2O_{8+y}$ with T_c = 84 K the lowest structural transition is at T_d = 95 K, while for BiSrCaCuPbO with $T_c \simeq 107$ K the respective anomaly is at $T_d \simeq 130$ K, much like the T_c versus T_d scaling in $La_{2-x}[Sr(Ba)]_xCuO_4$, $YBa_2Cu_3O_{7-y}$ discussed above and electrondoped cuprates[29].

1.5 Pseudogap

The pseudogap in the field of high-T_c superconductivity describes an energy which has very few states associated with it. It is similar to a 'gap' that an energy which has no allowed states. Such gaps open up, when electrons interact with the lattice. Several experimental techniques provide evidence for an unconventional normal state common to all of high-T_c superconductors and identified by the opening of a gap in the electronic excitation spectrum at a temperature T^* above the critical temperature T_c is called pseudogap (PG). This observation created an intense debate about its origin, since the answer to this question may turn out to be essential for the understanding of high-T_c superconductivity [32]. The experiments [33] by Li and coworkers pertain to the pseudogap region of the phase diagram of the cuprates (Fig.1.3). It is a sort of precursor state for discovering the physical principles that underlie the cuprates' behaviour. In the pseudogap region, at a temperature below T^* but above the temperature below which superconductivity emerges (T_c), all thermodynamic and electronictransport properties of cuprates change by a large quantity owing to the materials' loss of lowenergy electronic excitations.

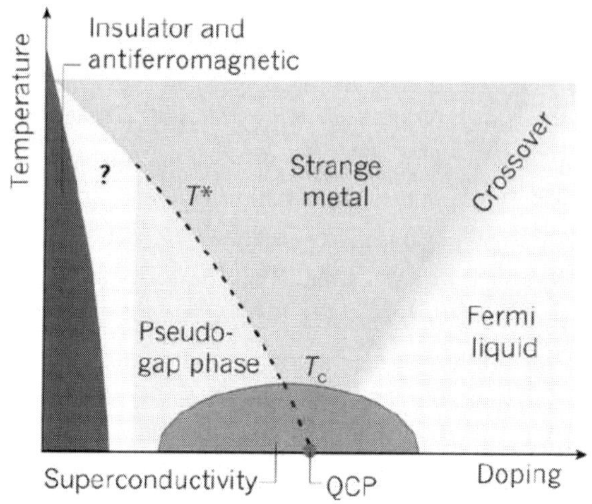

Fig. 1.3 Phase diagram of cuprates

The pseudogap region is bounded on one side by a region of remarkably simple but unusual properties that do not fit into the Fermi-liquid-type model which has been used to describe metals at low temperatures for a period of hundred years. Some researchers tried to understand this 'strange-metal' region early in the history of high-T_c superconductivity, by hypothesizing a quantum critical point in the dome-shaped superconductivity region of the phase diagram (Fig.1.3). This point would occur at zero temperature and would involve a change in the symmetry in electronic structure of the materials. Because T_c is determined by collective electronic excitations of the materials in the non-superconducting state, it is un arguable that the coupling of electrons to such excitations in the strange-metal region and their modifications in the pseudogap region lead to high-T_c superconductivity.

At very low levels of electron-hole doping, cuprates are insulating and antiferromagnetic. They become conducting with increasing doping levels, and the exact temperature and doping level determine which phase of matter they will be in. They become superconducting below T_c and above T_c but below T^* they fall into the pseudogap phase. The boundary of the pseudogap region at low doping levels is unknown. The transition between the Fermi-liquid phase and the strange-metal phase occurs gradually (by crossover). QCP corresponds to the quantum critical point at which the temperature T^* goes to absolute zero. Study of Li and coworkers [33] corresponds to the pseudogap phase.

1.6 Phase Diagrams of Some High-TcCuprates

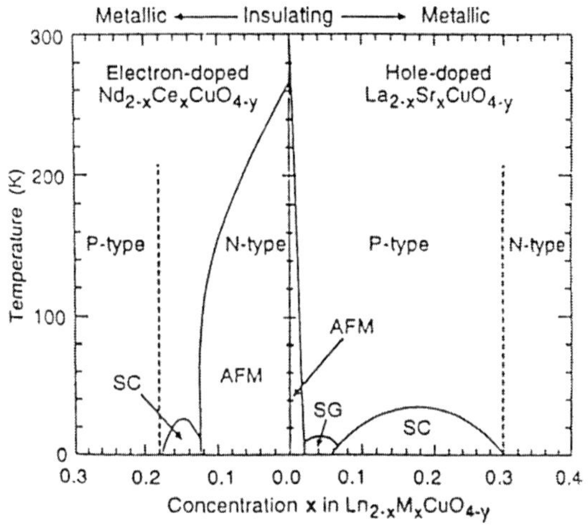

Fig. 1.4 Phase diagram of $Nd_{2-x}Ce_xCuO_4$ and La_{2-x} Sr_x CuO_4 (For comparision)

An actual phase diagram of LSCO is shown in fig.1.4 which is drawn temperature (T) vs Sr concentration x. For electron doped (NCCO) superconductors the antiferromagnetism (AFM) destroys at about 14% of doping but the superconductivity (SC) arises 12% of doping, hence the AFM and SC coexists. The T_c is maximum around 15% of doping and SC destroys around 18%. For hole doped (LSCO) superconductors the AFM destroys sharply about 4% doping and the superconductivity arises after 5% of doping. In between AFM andSC there is a spin glass phase. The T_c is maximum around 16% of doping. The SC exists up to around 29% of doping. Hence the SC is a long range order. On the onset of optical doping a huge range of pseudogap region exist. The pseudogap temperature decreases with increase of the doping and destroys completely around the optical doping. The pseudogap may be arise due to CDW, SDW and J-T distortion etc.

1.7 Low-Temperature Tunneling Spectroscopy

The electron tunneling spectroscopy [34] has become the technique of choice to probe the superconducting quasi-particle density of states. It is probably the most successful and sensitive way to measure the gap in the electronic excitation spectrum at the Fermi energy. We shall review the differential tunneling conductance spectra dI / dV measured on selected HTS, with the aim of identifying generic features in the electronic DOS related to their superconducting properties. We discuss in particular the gap structure at the Fermi energy, the background conductance, the dip-hump feature above the gap, the zero-bias conductance peak (ZBCP) at the Fermi energy, and the doping dependence and spatial in homogeneity of the tunneling spectra.

METHODOLOGY

In high-T_c systems, the oxygen atoms surrounding the central Cu-atom form an octahedron. The structural evidences show that the neighboring Cu-O octahedral are enlarged and contracted alternately. As a result Cu-atoms acquire disproportionate charges, because of the two different Cu-O bond lengths. This indicates the existence of strong electron-phonon interaction. The presence of the nested pieces of Fermi surface and the disproportionate charges of Cu-site stabilizes the system with Fermi surface instability by the formation of the charge density wave (CDW). Long range antiferromagnetic order, present in cuprates owes its origin to itinerant electrons in the system. The presence of the Fermi surface instability due to the existence of the nested pieces of Fermi surface results in the formation of a spin density wave (SDW). Based upon our mean-field model [35] the Hamiltonian is written as

$$H = \sum_{k\sigma} \epsilon_k \, c^\dagger_{k\sigma} c_{k\sigma} + \Delta_c \sum_{k\sigma} c^\dagger_{k\sigma} c_{k+Q\sigma}$$
$$+ \Delta_s \sum_{k\sigma} \sigma \, c^\dagger_{k\sigma} c_{k+Q\sigma} + \Delta \sum_{k} (c^\dagger_{k,\uparrow} c^\dagger_{-k,\downarrow} + c_{-k,\downarrow} c_{k,\uparrow})$$

(1)

The 1st term in the above Hamiltonian (H) represents the hopping of copper d-electrons between neighboring sites. Here $c^\dagger_{k\sigma}$ ($c_{k\sigma}$) is the creation (annihilation) operator of the conduction electrons of copper atoms. The hopping takes place between the adjacent sites of copper with the dispersion $\varepsilon_k = -2t_0(\cos k_x + \cos k_y)$, where t_0 is the nearest-neighbour hopping integral. The 2nd term describes the CDW interaction, where Δ_c is the CDW gap parameter which is defined as

$$\Delta_c = -V_1 \sum_{k\sigma} < c^{\dagger}_{k,\sigma} c_{k+Q,\sigma} >$$

(2)

where V_1 is attractive Coulomb interaction in presence of phonon coupling to the electron densities of the conduction band. The CDW interaction arises due to the Fermi surface instability due to nesting condition $\varepsilon_{k\pm Q} = -\varepsilon_k$ The 3rd term describes the longitudinal spin density wave (SDW) interaction. The SDW interaction is a type of antiferromagnetic interaction arising due to the Fermi surface nesting. The SDW interaction originates from the repulsive Coulomb interaction among the electrons. The longitudinal SDW gap parameter is defined as

$$\Delta_s = +V_2 \sum_{k\sigma} \sigma < c^{\dagger}_{k,\sigma} c_{k+Q,\sigma} >$$

(3)

Here V_2 is the repulsive Coulomb interaction and the spin, $\sigma = \uparrow(\downarrow)$ corresponding to up(down) spin configuration of the electrons.

The last term describes the s-wave BCS type mean-field superconducting interaction present in the conduction band. The SC gap Δ is defined as

$$\Delta = -V_0 \sum_{k} < c^{\dagger}_{k,\uparrow} c_{-k,\downarrow} >$$

(4)

Here V_0 is the momentum independent effective attractive Coulomb interaction, which exists for the interval $-\omega_c \leq \varepsilon_k \leq \omega_c$, where ω_c is the cut-off energy.

Calculation of electron Green's Functions

For the calculation of the CDW, SDW and SC order parameters, we have to calculate their respective correlation functions from their respective single particle Green's functions. We define four electron Green's functions $A_j(k, \omega)$ for j = 1 to 4 for up-spin configuration of the electron spins.

$$A_1(k,\omega) = << c_{k,\uparrow}; c^{\dagger}_{k,\uparrow} >>_{\omega} \; ; \; A_2(k,\omega) = << c^{\dagger}_{-k,\downarrow}; c^{\dagger}_{k,\uparrow} >>_{\omega}$$

$$A_3(k,\omega) = << c_{k+Q,\uparrow}; c^{\dagger}_{k,\uparrow} >>_{\omega} \; ; \; A_4(k,\omega) = << c^{\dagger}_{-(k+Q),\downarrow}; c^{\dagger}_{k,\uparrow} >>_{\omega}$$

(5)

When these Green's functions are evaluated by using Zubarev's [36] Green's functions technique, the four coupled electron Green's functions are found to be

$$A_1(k,\omega) = \frac{1}{2\pi}\left[\frac{(\omega+\epsilon_k)(\omega^2-\epsilon_{2k}^2)}{|D(\omega)|}\right]$$

$$A_2(k,\omega) = \frac{1}{2\pi}\left[\frac{\Delta(\omega^2-\epsilon_{2k}^2-2\Delta_s(\Delta_c-\Delta_s))}{|D(\omega)|}\right]$$

$$A_3(k,\omega) = \frac{1}{2\pi}\left[\frac{(\Delta_c+\Delta_s)(\omega^2-\epsilon_{2k}^2)+2\Delta^2\Delta_s}{|D_(\omega)|}\right]$$

$$A_4(k,\omega) = \frac{1}{\pi}\left[\frac{(\omega+\epsilon_k)\Delta_c\Delta_s}{|D(\omega)|}\right]$$

6)

where $\varepsilon^2_{2k} = \varepsilon^2_k + (\Delta_c - \Delta_s)^2$ and $|D(\omega)|$ in the denominator of these above Green's functions is given by

$$|D(\omega)| = \omega^4 - \omega^2 (E^2_{1k} + E^2_{2k}) + E^2_{1k}E^2_{2k} - 4\,\Delta^2_s\,\Delta^2_c$$

(7)

Here the terms E_{1k} and E_{2k} appearing in Eq. (7) can be expressed as

$$E_{1k}, E_{2k} = \sqrt{\epsilon_k^2 + \Delta_c^2 + (\Delta_s \pm \Delta)^2}$$

(8)

The four quasi-particle energy bands i.e: $\pm\omega_{1k}, \pm\omega_{2k}$ can be obtained from the poles of the Green's functions given in Eq. (6), which are expressed as

$$\omega_{1k}, \omega_{2k} = \sqrt{\epsilon_k^2 + (W \pm \Delta_s)^2}$$

(9)

Due to the interplay of CDW interaction and the superconductivity, the SC gap is renormalised to give an effective gap $W = \sqrt{\Delta^2 + \Delta_c^2}$. This shows that the CDW and SC gaps appear together and they influence each other.

Hence, it indicates that the CDW gap acts as a pseudogap. Sometimes it becomes difficult to interpret the tunneling conductance spectra, Raman spectra and the neutron scattering spectra due to the co-existence of the CDW and SC interactions in the systems. In addition to this, the SDW interaction again renormalises, the quasiparticle bands ω_{1k} and ω_{2k} by the presence of $\pm\Delta_s$ terms. We apply on our model calculation to interpret the scanning tunneling microscopy experimental data.

2.2 The CDW, SDW and SC gap equations

In order to study the interplay of the CDW, SDW and SC interactions in their co-existence phase in the copper oxide superconductors, one has to calculate the gap equations from the Green's functions. The CDW gap (Δc) is calculated from the Green's function $A_3(k, \omega)$ given in Eq. (6) by using the definition of CDW gap defined in Eq. (2). The integral form of the CDW gap is written as

$$\Delta_c = -g_1 \int d\epsilon_k \left[F_{11} + F_{12} \right]$$

(10)

where the CDW coupling is $g_1 = N(0)V_1$ with N(0) as the density of states of the conduction band. The function F_{1j} (for j=1 to 2) can be written as

$$F_{1j} = \frac{\Delta_c(\omega_{jk}^2 - E_{3k}^2)}{\omega_{jk}(\omega_{1k}^2 - \omega_{2k}^2)} \tanh(\frac{1}{2}\beta\omega_{jk})$$

(11)

Where $E_{3k}=(\varepsilon_k^2+\Delta_c^2-\Delta_s^2)^{1\backslash 2}$

The expression for the SDW gap is calculated from the correlation functions derived from the Green's function $A_3(k, \omega)$ given in Eq. (6) by using the definition of the SDW gap in Eq. (3). The integral equation for the SDW gap is written as

$$\Delta_s = -g_2 \int d\epsilon_k [F_{21} - F_{22}]$$

(12)

where the SDW coupling parameter is $N(0)V_0$. The functions F_{21} and F_{22} appearing in Eq. (12) are written as, the function F_{2j} for j = 1 to 2

$$F_{2j} = \frac{\Delta_s(\omega_{jk}^2 - E_{4k}^2 + 2\Delta^2)}{\omega_{jk}(\omega_{1k}^2 - \omega_{2k}^2)} \tanh(\frac{1}{2}\beta\omega_{jk})$$

(13)

Where $E_{4k} = (\varepsilon_k^2 - \Delta_c^2 + \Delta_s^2)^{1\backslash 2}$

The expression for SC gap is calculated from the correlation functions derived from the Green's function $A_2(k, \omega)$ shown in Eq. (6). The integral form of the gap equation for the superconducting order parameter is expressed as

$$\Delta = - g \int_{-\omega_c}^{\omega_c} d\epsilon_k \left[F_{31} + F_{32} \right]$$

(14)

where the superconducting coupling parameter $g = N(0)V_0$. Functions F_{3j} for j = 1 to 2 are written as

$$F_{3j} = \frac{\Delta(\omega_{jk}^2 - E_{3k}^2)}{\omega_{jk}} \tanh(\frac{1}{2}\beta\omega_{jk})$$

(15)

To corelate the strong energy dependence of the conduction electron density of
states around the center of the band in the system, we assume a model density of state $N(\varepsilon)$ as shown below

$$N(\epsilon) = N(0)(\sqrt{(1 - |\epsilon/D|)}) ln|D^2/\epsilon^2| \qquad (16)$$

where 2D is the conduction band width. This form of the density of state also mimics that for a tight binding band of a two dimensional square lattice, thus justifying the choice for the copper oxide system. The physical quantities involved are made dimensionless by dividing them by the hopping integral $2t_0$. The reduced parameters are, the superconducting parameter $z = \Delta/2t_0$; the CDW parameter $z_1 = \Delta_c/2t_0$; the SDW parameter $z_2 = \Delta_s/2t_0$;

the SC coupling $g = N(0)V_0$; the CDW coupling $g_1 = N(0)V_1$; the SDW coupling $g_2 = N(0)V_2$; reduced temperature $t = k_BT/2t_0$.

3. DISCUSSION

On the basis on our model, the gap equations for mutually competing orders i.e CDW, SDW and SC are calculated. The dimensionless CDW, SDW and SC parameters i.e. z_1, z_2 and z are involved integral functions and they are functions of each other and hence are solved self-consistently for a set of SC coupling g = 0.09994, the CDW coupling g_1 = 0.0445 and the SDW coupling g_2 = 0.0311. The temperature variations of the reduced SC parameter (z), CDW parameter (z_1) and the SDW parameter (z_2) are shown in Fig.3.1.

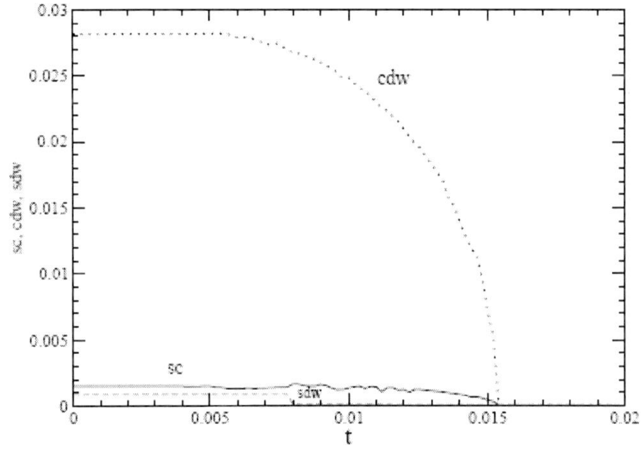

Fig. 3.1 The self-consistent plot of SC gap parameter(z) vs.temp.(t), CDW gapparameter(z_1) vs. t, and SDW gap parameter(z_2) vs. t for fixed values of g= 0.09994, g_1= 0.0445, g_2= 0.0311.

The SDW phasefor cuprates lies below the Ne´el temperature (t_N) $\simeq$ 0.008 and the magnitude of SDW gap is also much reduced. The SC and CDW phases co-exist below a superconducting transition temperature t_c $\simeq$0.016. Which corresponds to the SC transition temperature of T_c $\simeq$40K for the hopping integral $2t_0$=0.25 eV . This T_c $\simeq$40K is of the order of the SC transition temperature (T_c^{max} $\simeq$38K) for LSCO. It is observed the t_c and the CDW transition temperature t_d are equal. The pseudogap (CDW) merges with the superconducting gap showing that the superconducting gap is renormalised to give a higher gap value of $W=(\Delta^2+ \Delta_c^2)^{1/2}$ as observed from the relation for the quasi-particle band energy Eq. . Earlier panda et.al. [35] have solved the gap equations for the SC, CDW and SDW parameters for a set of higher coupling constants [i.e g=0.4512, g_1=0.0941 and g_2=0.1095] compared to the present set of couplings and obtained the temperature dependent plots for z, z_1 and z_2 with three distinctly different transition temperatures. Moreover, the temperature dependent CDW order parameter resembles the plot obtained for the Bismuth based superconductor by Ekino et. al. [37]. It is also noted that the pseudogap phase (CDW) for that system appears above the SC transition temperature (T_c^* = 86 − 89K) and below the pseudogap temperature (T_c $\simeq$120 − 130K). Such situation exists in the phase diagram of the under-doped high-T_c systems[38]. In the present computation for the dimensionless order parameters z, z_1 and z_2, we have taken smaller values of the three couplings and particularly very small value of SC coupling, g=0.09994. In this situation the SC and CDW transition temperatures are equal and, the magnitude of SC and CDW order parameters are also same order. Hence the superconducting order parameter is renormalized to give a higher magnitude of the SC gap i.e $W=(\Delta^2+ \Delta_c^2)^{1/2}$. At t_c=t_d, the CDW and SC phases coexist in the phase diagram in the under-doped regime of the high-T_c cuprates [38]. The effect of different couplings (i.e g, g_1, g_2) on the dimensionless parameters, z(t), $z_1(t)$ and $z_2(t)$ are shown in Figs. 3.2 and 3.3.

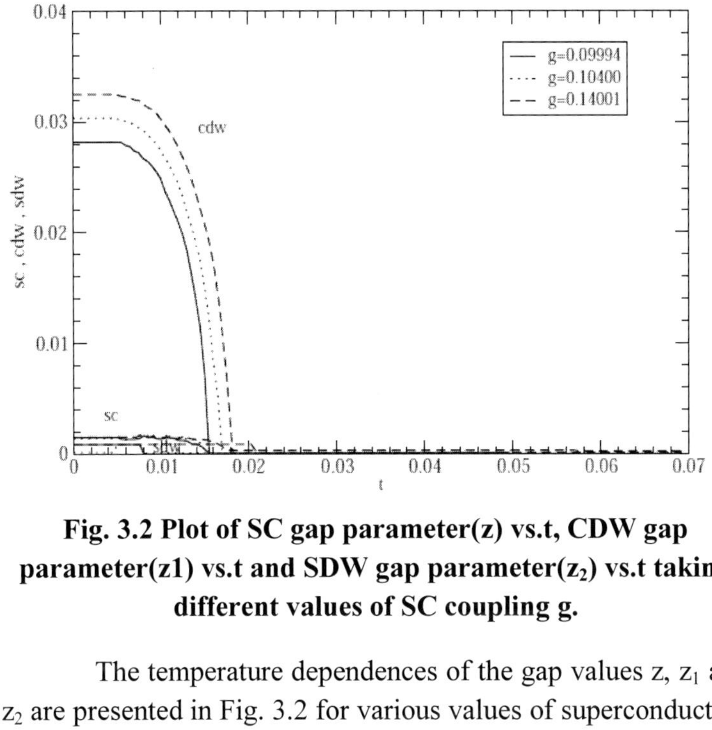

Fig. 3.2 Plot of SC gap parameter(z) vs.t, CDW gap parameter(z1) vs.t and SDW gap parameter(z_2) vs.t taking different values of SC coupling g.

The temperature dependences of the gap values z, z_1 and z_2 are presented in Fig. 3.2 for various values of superconducting coupling parameter (g). It is observed that with increase of SC coupling parameter, the magnitude of the SDW gap as well as the Ne´el temperature are suppressed resonably. Further with increase of SC coupling g, both the SC transition temperature (t_c) and the CDW transition temperature (t_d) are moved together to a higher values. Under these conditions, the magnitude of the CDW gap is increased through out the temperature range and, both the CDW and SC phases are extended to the higher temperature resulting in the increase of t_c due to the existence of the pseudogap (in this case CDW). This indicates strong interplay between the CDW and SC interactions. It is further noted that with increase of SC coupling (g), the CDW and the SC transition temperatures (t_c) are incresed very slowly, while the SDW transition temperature (t_n) is increased rapidly and the magnitude of the SDW gap remains at low costant value through-out the temperature range. In this calculation, the SC transition temperature t_c=0.16 ($T_c \simeq$40K for a hopping integral of $2t_0 \simeq$0.25 eV) for a given value of superconducting coupling g=0.14001. The SC transition temperature can still be enhanced by increasing the SC coupling. Thus we can achieve higher T_c even with with lower SC coupling as compared to BCS type metallic superconductors. Due to the pseudogap(CDW phase in the present case) interaction with superconductivity. In this calculation, we have considered BCS type weak SC coupling. Here it is noted that there exists controversy correspond to the electron-phonon pairing mechanism by Zhao[39,40] and the spin fluctuation pairing mechanism [41] for high-T_c cuprates. Plakida [41] has agreed that the spin fluctuation pairing mechanism supports the experimental observations with negligible contribution of electron-phonon interaction. On the other hand, Zhao [39,40] has observed that the strong electron-phonon interaction, rather than spin fluctuation, plays an important role in the microscopic pairing mechanism in cuprates. They further suggest that intrinsic pairing symmetry in the bulk is not d-wave, but extended s-wave in hole doped cuprates and node less s-wave in electron-doped cuprates.

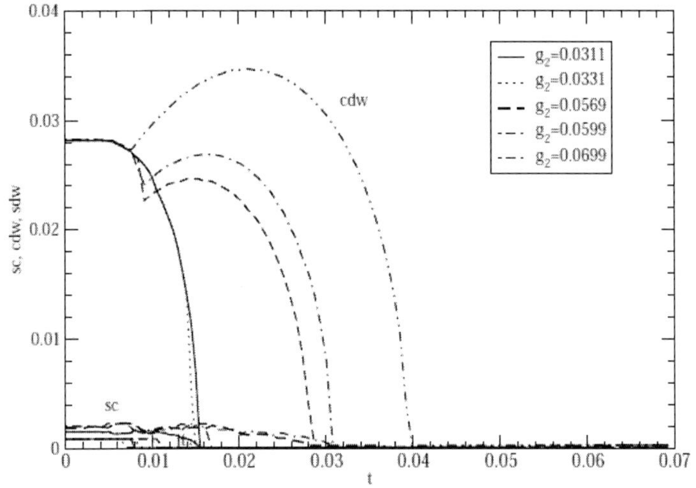

Fig. 3.3 Plot of SC gap parameter(z) vs.t, CDW gap parameter(z_1) vs.t and SDW gap parameter(z_2) vs.t,taking different values of SDWcoupling (g_2)=0.0311, 0.0331, 0.0569, 0.0599 and 0.0699.

The temperature dependences of z, z_1 and z_2 for different values of SDW coupling (g_2) are shown in Fig. 3.3. The SDW parameter is reduced with increase of SDW coupling and is finally suppressed completely. However, the effect of g_2 on CDW and SC gaps shows some new results. The CDW gap magnitude is enhanced considerably through-out the temperature range with the increase of SDW coupling. Under such condition, the SC transition temperature (t_c) and CDW transition temperature (t_d) are moved to new higher values and there by extending the superconducting phase to a higher temperatures. This shows that the SDW coupling plays an vital role in enhancing the SC transition temperature and the magnitude of the pseudogap (CDW).

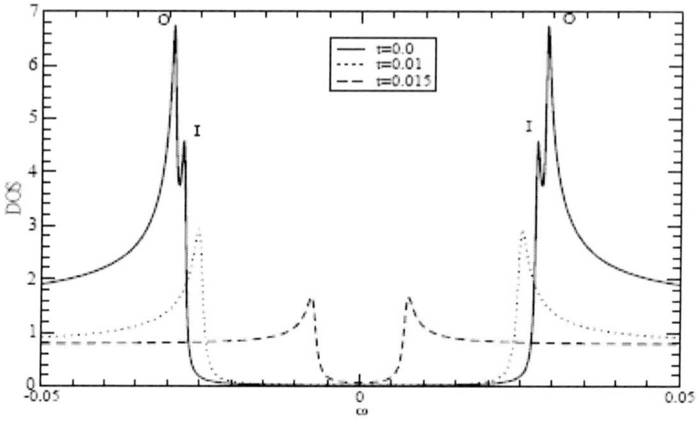

Fig. 3.4 Plot of the conduction electron DOS at temperature t= 0.0, 0.01, and 0.015.

The density of states (DOS) represents the tunneling conductance measured by STM technique. Thus, we have calculated the density of states from the spectral density function which is defined as $\rho_{k,\sigma}(\omega) = -2\pi \mathrm{Im}G(k, \omega)$ where G is the conduction electron Green's functions. The plot of DOS is shown in Fig.3.4 for several temperatures, t=0, 0.01, 0.015. The three phases such as SC, CDW and SDW co-exist at temperature t = 0. It is observed from Fig.3.4 that there appears two gaps at temperature t=0 i.e the outer gap, O-O with gap edges at $\pm(W+z_2) \simeq 0.0291$ and the inner gap, I-I with gap edges at $\pm(W-z_2) \simeq 0.0275$. These two gap edge values can be measured from the conductance data from the superconductor-insulator junction. Applying our model calculation, the reduced SDW gap z_2 and the renormalised SC gap, $W \simeq (z^2 + z_1^2)^{1/2}$ can be calculated. However, the individual magnitudes of the SC gap

(z) and the CDW gap (z_1) can not be calculated from our model calculation. The STM measurements [42] on the doped system LSCO with doping x=0 gives $\Delta_2 \simeq 5$meV and $\Delta_1 \simeq 10$meV. Our model calculation is applied to the system with $W + z_2 = 10$meV and $W - z_2 = 5$meV. This gives the renormalised SC gap of $W = (z^2 + z_1^2)^{1/2} = 7.5$ meV and the smaller SDW gap $z_2 = 2.5$ meV. The renormalised SC gap (W) obtained from any tunneling measurements is higher than the actual SC gap. This is in agreement with the remarks made by the authors [42]. They have observed that the magnitude of the superconducting gap is not in simple agreement with the SC transition temperatures. Our model calculation can be applied similar to the tunneling measurements on other cuprate systems [43-45]. For the temperature range $t_N < t < t_c$ i.e at the temperature 0.01, the SDW phase disappears, but the CDW and SC phases co-exist. Therefore, we observe a single gap at t=0.01 in the DOS with gap edges at $\pm W = (z^2 + z_1^2)^{1/2} = 0.0253$. Since the CDW and SC gaps are renormalised in the co-existence phase, it is not possible to find out the individual components of the two gap values. Thus, pseudogap (CDW) interferes with the SC gap, providing there-by an uncertainty in the measurement of the SC gap magnitude. Similarly a single gap structure with reduced gap size appears at temperature t = 0.015 in the vicinity of the SC transition temperature. The CDW state in cuprate systems acts as a pseudogap which exists in the normal state up to a temperature $t_d \simeq t_c$ as shown by the tunneling measurement for Bi-Sr-Ca-Cu-O system [37].

CONCLUSION

In the present work, we consider the strong interplay between the charge density wave, spin density wave and superconducting order parameters within a mean-field approximation. The expressions for these three order parameters are derived from the correlation functions calculated from the corresponding single particle electron Green's functions. The order parameters are solved self-consistently for a set of the CDW coupling g_1=0.0445 and the SDW coupling g_2=0.0311 and the SC coupling g=0.09994 satisfying the experimental condition, $\Delta_s < \Delta < \Delta_c$. In this model the SDW and CDW gap are assumed to act as pseudogaps. The pseudogap (CDW) gap merge with the superconducting gap, there-by giving rise to a renormalised superconducting gap $W= (z + z_c)^{1/2}$ which becomes higher than the pure superconducting gap. With the increase of SC coupling the SDW gap is suppressed, while the SC and CDW gaps are enhanced through-out the temperature range. It is to note further that both the SC and CDW transition temperatures are coincident and are shifted to higher value with the increase of SC coupling. This shows a strong interplay between the CDW (pseudogap) and the SC interactions. With the increase of the SDW coupling, the SDW gap is suppressed gradually, but the SC and CDW gaps are enhanced considerably.In this case also the SC and CDW transition temperatures shift together to new higher value. Thus it is clear that the SDW coupling plays an important role in the enhancement of the SC transition temperature. We have calculated the density of states (DOS) of the conduction electrons from the imaginary part of the Green's functions. The DOS is directly related to the tunneling conductance spectra obtained from the scanning tunneling microscopy. The DOS displays two gap structures i.e outer gap O-O with the gap edges at $\pm(W + z_2)$ and the inner gap, I-I with gap edges at $\pm(W - z_2)$ where the renormalised SC gap is $W = (z^2 + z_1^2)^{1/2}$. From the measured gap edge values of the tunneling spectra one can calculate independently the renormalised SC gap W and the SDW gap z_2. At temperature $t > t_N$ the CDW and the SC gaps co-exist. Since these two gaps merge together giving rise to a single renormalised SC gap, we obtain only a single gap structure with gap edge values at $\pm W = (z^2 + z_1^2)^{1/2}$ in the co-existance phase. The renormalised SC gap W is certainly higher than the true SC gap. This anomaly in the SC gap magnitude arises due to the pseudogap effect (in this case CDW gap).

REFERENCES

[1] H. K. Onnes, *Proc. Roy. Soc. Amsterdam* 27 (1911) 75; Leiden Comm. 142a.

[2] J. Bardeen, L. N. Cooper, J. R. Schrieffer, *Phys. Rev.* 106 (1957) 162.

[3] J. R. Gavalar, *Appl. Phys. Lett.* 23 (1973) 480.

[4] F. Steglich, J. Aarts, C. D. Bredl, W. Lieke, D. Meschede, W. Franz, H. Schafer, *Phys. Rev. Lett.* 43 (1979) 1892.

[5] H. R. Ott, H. Rudigier, Z. Fisk, J. L. Smith, *Phys. Rev. Lett.* 50 (1983) 1595.

[6] G. R. Stewart, Z. Fisk, J. O. Willis, J. L. Smith, *Phys. Rev. Lett.* 52 (1984) 679.

[7] T. Ishiguro, K. Yamaji, G. Saito, *Organic Superconductors,* 2nd ed., (Berlin:Springer) 1998.

[8] A. F. Hebard, M. J. Rosseinsky, R. C. Haddon, D. W. Murphy, S.H. Glarum, T. T. M. Palstra, A. P. Ramirez, A. R. Kortan, *Nature* 350 (1991) 600.

[9] K. Tanigaki, T. W. Ebbesen, S. Saito, J. Mizuki, J. S. Tsai, Y. Kubo, S. Kuroshima, *Nature* 352 (1991).

[10] J. G. Bednorz, K. A. M¨uller, *Z. Phys. B* 64 (1986) 189.

[11] M. K. Wu, J. R. Ashburn, C. J. Trong, P. H. Hir, R. L. Meng, L. Gao, Z. L. Huang, Y. Q. Wang, C. W. Chu, *Phys. Rev. Lett.* 58 (1987) 908.

[12] P. C. Canfield, S. L. Budko, *Annu. Rev. Condens. Matter Phys.* 1 (2010) 26.

[13] Y. Kamihara, T. Watanabe, M. Hirano, H. Hosono, *J. Am. Chem. Soc.* 130 (2008) 3296.

[14] H. Takahashi, K. Igawa, K. Arii, Y. Kamihara, M. Hirano, H. Hosono, *Nature* 453 (2008) 376.

[15] H. Takagi, S. Uchida, K. Katizawa and S. Tanaka, *Jpn. J. Appl. Phys. Lett.* 26(1987) L123.

[16] A. Schilling, M. Cantoni, J. D. Guo and H. R. Ott, *Nature* 363 (1993) 56.

[17] H. Maeda, Y. Tanaka, M. Fukutomi, T. Asano, *Jpn. J. Appl. Phys.* 27 (1988) L209.

[18] Z. Z. Sheng, A. M. Herman, *Nature* 322 (1988) 138.

[19] A. Schilling, M. Cantoni, J. D. Guo, H. R. Ott, *Nature* 363 (1993) 56.

[20] C. W. Chu, L. Gao, F. Chen, Z. J. Huang, R. L. Meng, Y. Y. Xue, *Nature* 365 (1993) 323.

[21] A. Mourachkine, *Europhys. Lett.* 50 (2000) 663.

[22] K. Yamada, K. Kurahashi, T. Uefuji, M. Fujita, S. Park, S.-H. Lee, Y. Endoh, *Phys. Rev. Lett.* 90 (2003) 137004.

[23] P. Fulde, *J. Low Temp. Phys.* 95 (1994) 45.

[24] G. Shirana, Y. Endlh, R. J. Birgenau, M. A. Kastner, Y. Hideka , M. Oda, M. Suzuki and T. Murkamini, *Phys. Rev. Lett.* 59 (1987) 1613 .

[25] M. Mastuda et al., *Phys. Rev. B* 42 (1998) 10098 .

[26] J. M. Tranquada, D. E. Cox et al., *Phys.Rev. Lett.* 60 (1987) 159.

[27] Y. Tokura , H. Tagaki and S. Uchida, *Nature* 337 (1989) 345 .

[28] J. D. Jorgensen, Advances in Superconductivity XII (Proc. of the 12th Int. Symposium on Superconductivity (ISS*99) October 1719 (1999) Morioka) ed T. Yamashita and K. Tanabe (Tokyo: Springer) 9 (2000).

[29] M. Lang, R. K¨ursch, A. Grauel et al., Phys. Rev. Lett. 69 (1992) 482.

[30] J. M. Tranquada, *Physica B* 241-243 (1998) 745.

[31] Y. Koyama, Y. Wakabayashi, K. Ito and Y.Inoue, *Phys. Rev. B* 51 (1995) 9045.

[32] T. Timusk and B. Statt, *Rep. Prog. Phys.* 62 (1999) 61.

[33] Y. Li et al., *Nature* 468 (2010) 283.

[34] I. Giaever, *Phys. Rev. Lett.* 5 (1960) 147.

[35] S. K. Panda and G. C. Rout, *Physica C* 469 (2009) 702.

[36] D.N. Zubarev, Sov. Phys. Usp. 3(1960) 320.

[37] T. Ekino, Y. Sezaki, H. Fujji, *Phys. Rev. B* 60 (1999) 6916.

[38] Peter Wahl, *Nature Phys.* 8 (2012) 514.

[39] Guo-meng Zhao, *Phys. Scr.* 83 (2011) 038302.

[40] Guo-meng Zhao, *Phys. Scr.* 83 (2011) 038304.

[41] N. M. Plakida, *Phys. Scr.* 83 (2011) 038303.

[42] T. Kato, T. Maruyama, S. Okitsu, and H. Sakata, *J. Phys. Soc. Jpn.* 77 (2008) 054710.

[43] T. Yoshida, M. Hashimoto, I. M. Vishik, Z. Shen and A. Fujimori, *J. Phys. Soc. Jpn.* 81 (2012) 011006.

[44] K. Tanaka, W. S. Lee et.al., *Science* 314 (2006) 1910.

[45] K. L. Mohanta, S. K. Panda, B. K. Raj and G. C. Rout, *Phys. Express* 3 (2013) 29(1-6).

CHAPTER FOUR

ANTIFERTILITY ACTIVITY OF SIDA ACUTA BURM. F

Dusmanta Kumar Pradhan

Apollo College of Pharmacy,Anjora, Durg, CG

dkpradhan74@gmail.com

ABSTRACT

Herbs and their extracts/fractions are used for the treatment of different diseases and as antifertility agents. Different plants have been used as antifertility agents in the folkloric medicine. There is no scientific validation for the use of these plants. Wehave selected root of Sida acuta Burm. f. (Malvaceae) to prove its antifertility activity based on traditional claims/practices.The plant was identified by the Botanical Survey of India, Central National Herbarium Howrah, Kolkata, India. The plants were collected and roots were sepatrated from aerial parts. Roots were shade dried, powdered and extracted with pet. ether and methanol succesively. Extracts were subjected to preliminary phytochemical and chromatographic studies. The preliminary phytochemical investigation indicated the presence of alkaloids, phenolic compounds tannins, flavonoids, sterols, carbohydrates, glycosides, proteins and amino acids.The methanolic extract showed higher extractive value as comapred to pet. ether extract and for that reason the methanolic extract was selected for futher studies. The methanolic extract was screened for its antifertility activity using post coital antiimplantation, abrortificial, sperm motility (in vitro), estrogenic and antiestrogenic models. The methanolic extract of Sida acuta exhibited potent post coital anti-implantation, abortion, estrogenic and antiestrogenic activity and reduction in sperm motility. Thus the scientific validation for the traditional claims for antifertility use of plant has been justified.

Keywords: Antifertility, Alkaloid, antiimplantation, antiestrogenic activity.

4.1 INTRODUCTION

The species of Sida (Malvaceae) are annual herbs or under shrubs, mostly distributed in tropics and sub tropics of the world. The genus Sida comprises of about 200 species in the world, out of which twelve are occurring in India [1]. Most of the Sida species found in India are known by general name "Bala" and referred under "Bala Chatustaya" in Ayurvedic system of medicine. The drug Bala has been mentioned in Charak Samhita, under Brihaniya Kasaya, by the name of Vatyayani, under Prajasthapan Mahakasaya by the name of Vatyapushpi. It is also found mentioned under Balya Mahakashaya and Madhurskandha and in Sushrut Samhita under Vatasanshaman Verg under the name of Bala. The most medicinally important Indian species of Sida are S. acuta, Sida cordifolia, S. rhombifolia, S. spinosa (S. alba), S. cordata, S. javensis, S. mysorensis and S. ovata.

S.acuta is a weed, which is frequently found in pastures, wastelands, cultivated lands, roadsides, lawns, and in planted forests [2]. Once the plant becomes established, it is very competitive, holding and denying sites to other plants. It does appear to do best in disturbed habitats. The plant grows well in many soils, including some heavy clay and can tolerate dry as well as high rainfall conditions. This is more commonly found in Odisha and adjoining states.

Sida acuta Burm.f. a small, erect, much branched, annual shrub or herb; ranging from 30 to 100 cm in height, with a strong taproot; stem and branches flattened at the extremities, fibrous, almost woody at times. Branchlets pilose or subglabrous. Leaves ± distichous; stipules filiform , 4-6 mm, often longer than petiole,usually persistent ; petiole 4-6 mm, sparsely pilose; leaf blade ovate , oblong , lanceolate, or linear-lanceolate, 2-5 × 0.4-1 cm, both surfaces glabrous or sparsely stellate pilose, rarely with simple hairs adaxially, base obtuse , margin dentate , sometimes partly entire toward its base, apex acute or acuminate. Flowers solitary or paired, axillary sometimes congested at stem apex. Pedicel 4-12 mm, pilose, articulate at middle. Calyx shallowly cup-shaped, connate in basal 1/2, ca. 6 mm, mostly glabrous, margins often ciliate, lobes 5, caudate. Corolla yellow, less often white or yellow-orange, 8-10 mm in diam.; petals obovate, 6-7 mm, ciliate, base attenuate, apex rounded. Filament tube ca. 4 mm, sparsely hirsute. Schizocarp nearly globose; mericarps (4-) 6(-9), segmentiform-tetrahedral, ca. 3.5 mm, basally transversely ridged, side walls reticulate-veined, glabrous, apex beaked , ± extending into 2 awns, apically dehiscent . Seed trigonous, ca. 2 mm, glabrous except around hilum.

4.2 METHODOLOGY

4.2.1 Materials & Methods:

a) Preparation of extract solution:

The methanolic extract of ***Sida acuta*** was dissolved in Tween 80 (1%) in normal saline and the later were used for administration.

b) Selection of Animals:

Healthy adult female and malealbino rats of wistar strain weighing 150-200 g for antifertility andimmature female rats of 21-23 days old (wistar strain) weighing 40-60g and adult albino mice of swiss strain weighing 20-25 g of either sex wereselected for toxicological studies.

All the animals were obtained from Dept. of Pharmaceutical Sciences, BIT, Mesra, Ranchi, for the studies. They were housed under standard condition of temperature (26±2°C), relative humidity (45-55%), light and dark cycle (14:10 h) and fed with standard pellet food and water *ad libitum*. The initial body weight of each animal was recorded. The vaginal smear of the female rats was studied microscopically for estrus cycle every morning at 8-9 am. Only female rats with normal estrus cycle were selected for the antifertility activity evaluation. All experimental procedures were carried out in strict accordance with the guidelines prescribed by the Committee for the Purpose of Control and Supervision on Experimentation on Animals (Regd.No: 621/02/ac/CPCSEA) Provisional Protocol Approval No. BIT /PH/ IEC/08 /2011 [3].

Acute Toxicity of *Sida acuta*:

Administration of doses:

The acute toxicity study was performed as described by Turner [4]. Adult albino mice of either sex weredivided into 5 groups containing six animals ineach group. The mice were fasted for 18 h with water *ad libitum*. The extract suspension was administered orally at four different doses of 1000, 2000, 4000 and 6000 mg/kg body weight, respectively todifferent groups. Control received the vehicle (Tween-80, 1%, p.o.) only.

Animals were observed at regular time intervals at least once at every 30 minutes of initial after dosing, for the first 24 hrs. Additional observations like behavioral changes change in skin, fur, eyes, mucous membranes, respiratory, circulatory, autonomic and central nervous systems and somato motor activity, behaviour pattern and mortality was observed up to 72 hrs. Attention was also given to observation of tremors and convulsions.

ANTIFERTILITY ACTIVITY OF *SIDA ACUTA*

Post-coital Antifertility Testing

Vaginal smears from each rat were monitored daily, only rats with normal estrous cycle were selected. Anti-implantation activity was determined as described by Khanna and Chaudhary [5]. Beginning at 77 days of age, the female rats were paired according to numerical sequence with one untreated male (this being approximately of the same age, selected from the same batch and kept under the same conditions as female).Mating was monogamous (one male to one female). Every morning during the mating period, each female was examined for the presence of sperm plugs or sperm in vaginal smears. Females not showing a vaginal plug or sperm were returned to the male. This procedure was continued for approximately 2 weeks.The plant extract wasadministered at 200 and 400 mg/kg bw orally from days 1 to 7 of pregnancy. Control rats received the vehicle (Tween80, 1% p.o.) The animals were laparotomised under light ether anaesthesia and semi-sterile condition on day 10 of pregnancy. Both horns of the uterus were observed for the number and size of implants. The rats were allowed to recover and were allowed to give birth. The gestation period was taken as the time between the day of successful mating and commencement of the birth (i.e. first detected presence of offspring, post partum day 0). During gestation and postpartum period the females were individually caged. The total litter size was counted as soon as possible after parturition. On day 1 post partum, live pups were sexed, weighted and examined for external abnormalities. The litters were allowed to grow to check their postnatal growth and monitor any congenital abnormalities up to 7 days postpartum period [5-8].

Abortifacient Activity:

Female albino rats (150 -200 g) of the normal estrous cycles were maintained and provided with food and water *ad libitum*. The females of proven fertility were left overnight with proven fertility male albino rats with 2:1 ratio at early estrous stage of estrous cycle. Then vaginal smear of female was observed for the presence of the sperms and formation of vaginal plug. Subsequent day was selected as day one of pregnancy. Such pregnant females were grouped into three groups of six each.

Group I: Served as positive control, received vehicle (1% Tween-80 in saline water) orally for one to fourteen days of pregnancy.

Group II: Received 200 mg/kg bw suspension of plant extract for one to fourteen days of pregnancy.

Group III: Received 400 mg/kg bw suspension of plant extract for one to fourteen days of pregnancy.

On 10th day of pregnancy, laparotomy was performed under light ether anesthesia and semi sterile conditions to know the presence of implantation sites in the uterine horns. The abdominal cavities are sutured and rats were allowed to continue the pregnancy to the full term. Second, laprotomy was performed three days after completion of treatment i.e. on the 18th day of pregnancy the number implantation sites counted and compared with initial number of implantation on 10th day of pregnancy.The percentage of pregnancy failure among treated groups was calculated [9].

Estrogenic and Antiestrogenic Activity:

The uterine weight and vaginal cornification method was employed for this assay [10]. Immature ovariectomised female albino rats (Wistar strain), 21-23 days old and weighing between 40 to 60 g were used. They were divided intosix groups consisting of six animals in each group. The first group served as control and received the vehicle only (Tween-80, 1%). The second group received a suspension of ethinyl estradiol in distilled water using Tween-80, 1 % at a dose of 0.03 mg/kg bw. The third and fourth groups received the methanolic extract at dose level of 200 and 400 mg/kg bw respectively and the fifth and sixth groups received, in addition to 0.03 mg/kg bw of ethinyl estradiol, a test dose of the methanolic extract at 200 and 400 mg/kg bw respectively. All the above treatments were given orally for 7 days. Body weights and vaginal smears were recorded daily. Positive smears were those containing nucleated or cornified cells and not more than a few leucocytes. On the 8th day of the experiment, all the animals were sacrificed by decapitation under light ether anesthesia and the uteri were dissected out, surrounding tissues removed, blotted on filter paper and weighed quickly on a sensitive balance.

Sperm motility (in-vitro study):

Six healthy adult male albino Wistar rats (150 -200 g) were used for the study. The animals were provided with foodand water *ad libitum*. The rats were anesthetized using ether anesthesia. The skin of scrotum was cut open tunicavaginalis. The small incision was made to make a rent so that epididymus can come out. After pulling out the testes, contents of caudaepididymus were removed and flushed in 1 ml of previously warmed buffer saline at 37°C to make the suspension of spermatozoa. The testes was returned back to the scrotal sac and wound was sutured properly [11]. At least 80 % initial normal sperm motility and 2.5 X 10^7/cc sperm count was considered for the selection of the sample. Wet drop technique was applied to study the motility of spermatozoa [12]. 10 μl of sperm suspension and 10 μl of methanolic extract at different concentrations (1.0, 2.5, 5.0 and 10.0 %) were placed on clean and dry glass cavity slides. Then it was mixed uniformly and examined under binocular research microscope at 10X and 40X magnifications. The motility of sperm was observed at various time intervals from 0, 15, 30, 60, 90, 120 and 150 seconds. In control, 10 μl of saline buffer solution was used. All values were reported as mean ± SEM.

StatisticalAnalysis

Pharmacological data were analyzed by one-way Analysis of Variance (ANOVA) followed by Dennett's t *test* with equal sample size (n = 6). The difference was considered significant when p value < 0.05 and 0.01. All the values were expressed as mean ± standard error mean (S.E.M.).

RESULTS:

Table 1: Post-coital antifertility activity of MESA root

Treatment	Dose, mg/kg body weight	No.of corporaleutia	No.of implantation sites	No. of rats having no implantation sites on day 10/ No. used	No. of implantation Activity(%)	Total litter size (mean ± S.E.M)
Control	-	8.5±0.44	10.0±0.43	0/6	0	7.34±052
MESA	200	8.0±0.36	8.35±0.67*	2/6	33.33	4.52±0.31**
MESA	400	7.67±0.31	6.54±0.51**	3/6	50.00	3.29±0.62**

Values are expressed as Mean ± S.E.M., n=6. *p < 0.05, **p < 0.01

Table 2 Offspring data from female Wistar rats treated with MESA root.

Treat-ment (doge, mg/kg body weight)	At birth		On day 1				
	Live litter size (mean ± S.E.M)	LBI (%) (mean ± S.E.M)	PL (%) (mean ± S.E.M)	Live litter size (mean ± S.E.M)	CPL (%)	Litter weight (g) (mean ±S.E.M.)	Mean pup weight (g) (mean± S.E.M.)
Control	9.0±0.17	95.29±2.82	1.93±2.15	9.0±0.17	0.0	61.5±0.	7.15±0.

						29	20
MESA (200)	5.3±0 .4	92.5±1.6 9^{ns}	1.89±2 .7^{ns}	5.3±0. 4	0. 0	42.13±0 .77	7.27±0. 12
MESA (400)	1.7±0 .3	81.943± 5.6^{ns}	2.02±3 .1^{ns}	1.7±0. 3	0. 0	18.53±0 .94	7.2±0.0 6

Values are expressed as Mean ± S.E.M. n = 6.$^{ns}p > 0.05$.

Table 3 Offspring data from female Wistar rats treated with MESA root

Treat-ment (doge, mg/kg body weight)	On day 4				On day 7				
	Live litter size(mean ± S.E.M)	CPL (%)	Litter weight (g)(mean ±S.E.M.)	Mean pup weight (g) (mean±SEM.)	Live litter size (mean ± S.E.M)	CPL (%)	Litter weight (g)(mean ±S.E.M.)	Mean pup weight (g) (mean± S.E.M.)	Viab-ility Index (%)
Control	9.0±0.17	0.0	98.53±1.04	11.03±0.18	9.0±0.17	0.0	140.37±1.01	15.60±0.46	100
MESA (200)	5.3±0.4	0.0	58.47±1.16	11.67±0.27	5.3±0.4	0.0	75.83±0.7	15.03±0.20	100
MESA (400)	1.7±0.3	0.0	24.37±2.3	11.10±0.26	1.7±0.3	0.0	34.53±1.04	14.67±0.09	100

Values are expressed as Mean ± S.E.M. n = 6.

Formulas for calculation:

LBI (Live birth index) = Pups born alive/total number of pups x 100.

PL (Pup loss)= Total litter size at birth-live litter size/total litter size x100.

CPL (Cumulative pup loss) = live litter size at birth – live litter size / live litter size at birth x 100.

Table 4 Abortifacient activity of MESA root

Group	Treatment	No of rats pregnant	No of rat implantation on 10th day	No of rat implantation on 18th day	% of Abortion
1	Control	06/06	10.0±0.57	10.0±0.57	00
2	MESA 200mg	06/06	7.0±1.0	6.0±1.0	40
3	MESA 400mg	06/06	6.0±1.52	5.0±1.15	50

Values are expressed as Mean ± S.E.M.

Table 5 Estrogenic and antiestrogenic activity of MESA root in female rats

Treatment Groups	Dose (mg/kg)	Uterine weight mg/100g body weight	Vaginal cornification
Control	Tween-80, 1%	82.00 ± 1.06	Nil
Ethinylestradiol	0.03	371.16 ± 3.01**	+++
MESA	200	93.50 ± 1.83*	+
MESA	400	120.66 ± 0.88*	++
Ethinylestradiol + MESA	0.03 +200	320.83 ± 1.74**	+++
Ethinylestradiol + MESA	0.03 +400	265.16± 1.66**	+++

Values are expressed as Mean ± S.E.M., n = 6. *p < 0.05, **p < 0.01 vs. control. +: nucleated epithelial cells; ++: nucleated and cornified cells; +++: cornified cells.

Table 6 Effect of MESA root on motility of rat spermatozoa (in vitro) percent motility of Spermatozoa at different duration (Sec)

Treatment	0 Sec.	15 Sec.	30 Sec.	60 Sec.	90 Sec.	120 Sec.	150 Sec.
Control	80.83±2.12	81.83±1.79	81.33±1.58	81.5±1.97	81.33±1.16	81.00±1.65	81.00±1.67
MESA 1%	80.50±1.02	56.50±1.45*	49.50±0.76*	41.83±1.10*	32.16±1.42*	11.66±0.66*	Motility nil
MESA 2.5%	80.83±0.79	50.16±1.16**	38.33±0.95**	27.00±0.96**	6.5±0.42**	Motility nil	
MESA 5%	80.83±0.98	35.66±0.55**	16.00±0.68**	6±0.36**	Motility nil		
MESA 10%	80.50±0.92	17.16±0.60**	7.5±0.42**	Motility nil			

Values are expressed as Mean ± S.E.M. n = 6. *p < 0.05, **p < 0.01

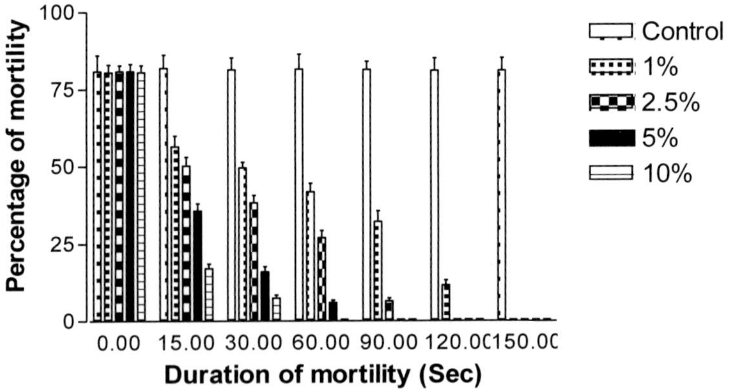

Figure 4.1 Effect of methanolic extract of sida acuta root on motility of rat spermatozoa in vitro. percent motility of spermatozoa at different duration (sec).

Acute Toxicity:

No mortality and changes in the behaviour wereobserved in all the treated and control groups of mice up to a dose of 2000 mg/kg body weight.At higher doses, there was aregular dose-dependent increase in mortality in mice. The firstmortality latency was 72 to 120 h after oral administration of 4000 mg/kg dose of extract and all animals died at the highest doseof 6000mg/kg p.o. In addition to death, this dose of plant extractalso induced minor to accentuated hypoactivity, ataxiaand lethargy. It was found that all the animals were safe at a dose of 2000 mg/kg bw and there was no abnormal behavior. Therefore, we have selected the $1/10^{th}$ and $1/5^{th}$ tolerated dose i.e. 200 mg/kg bw and 400 mg/kg bw as a therapeutic dose for antifertility studies.

Post coital antifertility activity:

The anti-implantation activity is expressed as the percentage of animals showing absence of implantations in uteri when laparotomised on day 10 of pregnancy. (Table 1).

The methanolic extract of *S. acuta* root exhibited significant ($p < 0.01$) anti-implantation activity in a dose dependent manner. All treatments reduced the number of litters born (total litter size) significantly ($p < 0.01$) confirming the antifertility activity of the plant used. This may be due to abortion. However, no vaginal bleeding was observed. Some rats treated with the extract at 400mg /kg bw, did not deliver any litters.

No toxic effects were observed either by gross visual examination or in the weight of animals. After discontinuation of treatment, all the animals were mated. This resulted in pregnancy and deliver of normal litters, indicating that the action of the extract was reversible. Hence, it revealed highest antifertility (anti-implantation as well as abortificient) activity.

Litter data:

Offspring data showed no statistical differences between groups in pup loss, cumulative pup loss and mean pup weight throughout the 7-day postpartum period. Similarly, no differences between groups were seen in live birth and viability index (Tables 2 and 3).

Abortifacient activity:

Pregnant rats receiving 200 and 400 mg/kg bw of methanolic extract of *S.acuta* root exhibited 40 % and 50 % abortion rate respectively as compared to control (Table 4).

Estrogenic and antiestrogenic activity:

The effect of the methanolic extract of *S. acuta* root on the immature rat uterus is shown in Table 5. Oral administration of the extract (200 and 400 mg/kg bw) caused significant increase in uterine weight in immature ovariectomised rats (versus control $P < 0.01$). The uterotrophic potency was less than that of ethinylestradiol. The methanolic extract at their contraceptive dose level significantly increased uterotrophic response when compared to control rats. The extract at the two doses of body weight induced vaginal opening and smear showed proestrous or estrous. The number of cornified cells in vaginal smear was considerably higher (+ to ++) than that of the control, but notably less than that of ethinylestradiol (+++). Simultaneous administration of extract caused a significant increase in uterine weight ($P < 0.01$), but the extent of the uterotrophic response was less than that produced by ethinylestradiol alone. The extract therefore has weak estrogenic activity at their contraceptive dose level when given alone. However, along with ethinylestradiol, they exhibited strong antiestrogenic property.

Motility of rat spermatozoa (*in vitro* study):

Initial motility of control sperm was 80.83 %, which remained almost static (81.0 %) up to 150 seconds as observed in the present study (Table 6). But sperm motility appears to decrease with time and increase with dose of methanolic extract. Dose dependent response of methanolic extract was noted *in vitro* motility of sperm in rat.

In this study, sperm motility was observed to be practically nil at 10 % concentration for methanolic extract in 60 sec. While, 5 % of the same showed nil in 90 sec, 2.5 % in 120 sec and 1 % in 150 sec (Table 6).

DISCUSSION

The Post coital antifertility effects observed in the present study could be due to one or more of the active constituents present in the methanolic extractof S.acuta root. In fact, it is well known that estrogenic substances inhibit pregnancy by suppressing the level of both follicular stimulating hormone (FSH) and luteinizing hormone (LH), which in turn prevent the implantation. The extract showed dose dependent anti-implantation activity. In higher dose animals were not showed any litter born, it revealed highest antifertility activity.Abortion refers to the premature expulsion of the products of conception from the uterus. It may be due to maternal exposure to phytochemicals, which can disrupt pregnancy and cause detachment of the embryo. The extract has shown significant abortifacient activity. The motility was seen in dose dependent manner.

It is well known that for implantation the exact equilibrium of estrogen and progesterone is essential and any disturbance in the level of these hormones may cause infertility [13]. The compound with hormonal activity usually disturbs the hormonal milieu in the uterus and provokes an infertility effect in experimental animals. In this study, the anti-implantation activity may be due to estrogenic activity, causing the expulsion of the ova from the tube, disrupting the luteotrophic activity of the blastocyst [14,15].

In another set of experiment on immature ovariectomized rats, we observed that the extract exhibited estrogenic activity as shown by the significant increase in uterine weight and vaginal cornification in immature female rats as compared to controls.The extract acted as a weak estrogen when given alone, but when given along with ethinyl estradiol, it exhibited strong antiestrogenic nature, since antiestrogen substances decrease the weight of the uterus [16]. This indicated that methanolic extract of S. acuta is a competitive antagonist to ethinyl estradiol.

Use of chemical contraceptives by women is effective but has some limitations owing to some undesired effects. Since last decade, efforts are being made to work on other areas such as spermicidal agents [17,18]. Herbal preparations have also been tested for spermicidal action in number of countries. In earlier studies it has been observed that alkaloid such as chloroquine, quinine and arecoline had the strongest reduction effect on human sperm motility and the effect was strongly dose dependent and resists the process of spermatogenesis and decrease sperm motility in rats [19-21].In the present study, our findings clearly indicated that whenever there was an increase in the dose of methanolic extract of S .acuta, there was a decrease in the complete immobility time of sperm (Table 4-21). The findings of the present study clearly indicated that the methanolic extract of root of S. acuta possesses antimotility action on spermatozoa in rats, may be due to its alkaloidal content.

The previous literatures reported Sida acuta contains triterpene □-amyrin, vomifoliol, □-sitosterol (sterol), Dimethyl terepthalate tocopherol derivative, 7a-methoxy-α-tocopherol, taraxastane triterpene, taraxast-1, 20(30)-dien-3-one, β-tocopherol, α-tocopherol, α-tocospiro B, taraxasterone, evofolin-A, and evofolin-B (Phenolic compounds) [22-24], N-trans-feruloyltyramine (Phenylpropanoids) [25], quercetin, kaempferol , apigenin, myricetin and genistein (Flavanoids), 5,7-dimethoxycoumarin (Coumarin) [26], and alkaloids ephedrine, swainsonine, cryptolepine and quindoline Heraclenol, acanthoside B quindolinone, cryptolepinone, 11-methoxyquindoline, isoquinoline alkaloids, indole alkaloids [27].

Alkaloid Vasicine, isolated from Adathoda vasica showed potent abortifacient and uterotonic effects in guinea pigs [28]. Flavonoids isolated from Striga lutea and Striga orobanchioides possessed strong estrogenic and antifertility properties [29]. Phytochemical studies indicated the presence of alkaloids, flavonoids, sterols and glycosides in MESA root. Since several of these compounds are known to exhibit antifertility activity [29-36], the antifertility effect of MESA might be due to the presence of one of these compounds.

4.3 CONCLUSION

In conclusion, the methanolic extract of S. acuta exhibited potent anti-implantation, abortion, estrogenic, antiestrogenic and reduction in sperm motility. Thus the scientific validation for the traditional claims for antifertility use of plants is justified. Further studies are needed to explore the possible mechanism of claimed actions of this plant and isolation of the individual compounds confirming the pharmacological activity.

REFERENCES

[1] Sivarajan VV, Pradeep KA (1996). Malvaceae of Southern Peninsular India: A taxonomic monograph Daya Publishing House, New Delhi.

[2] Mann, A, Gbate M, Umar AN (2003). *Sida acuta* subspecie *acuta.*Medicinal and economic palnt of Nupeland. Jube Evans Books and Publication 241.

[3] Ecobichon DJ (1997). The basis of toxicology testing, CRC Press, New York pp 43-49.

[4] Turner RA (1971). Screening Methods in Pharmacology, vol. II Academic Press, New York.

[5] Khanna U, Chaudhury RR (1968). Antifertility screening of plants. Part I, Investigation of Butea monosperma (Lam) Kutze, Indian J Med Res 56: 1575-1579.

[6] Badami S, Aneesh R., Sankar S, Sathishkumar MN, Suresh B,Rajan S (2003). Antifertility activity of *Derris brevipes* variety coriacea. Journal of Ethnopharmacology 84: 99-104.

[7] Dixon Rl, Hall JL (1982). Reproductive Toxicology. In: Hayes A.W.(Ed), Principles and Methods of Toxicology. New York Raven press pp 107-39.

[8] Long JA, Evans HM (1992). On the estrous cycle in the rat and its associated phenomena.Berkeley, CA: University of California Press.

[9] Mukharjee P (2001). Quality Control Methods of Herbal Drugs", 1 ed. Kolkota: Business Horizon Pharmaceutical pp 578-580.

[10] Edgren RA, Calhoun DW (1957). Chemical control of fertility. In: Edgren, R.A. (Ed.), Contraception.Marcel Dekker Inc, New York pp 537.

[11] Bhargava V, Prakash AO (2000). Spermicidal activity *in vitro* of bark extract of *Azadirachta indica* in rats. Ancient Sci. Life 20: 77-81.

[12] Sander FV, Cramer SD (1941). A practical method for testing the spermicidal action of chemical contraceptives. Human Fert pp 134-137.

[13] Psychoyos A (1966). Recent research on egg implantation. CIBA Found Study Groups 23.

[14] Pincus G (1965). Control of fertility. New York: Academic press.

[15] Anderson LL, Maghissi KS, Hafez ESE (1972). Biology of mammalian fertilization and implantation. Thomas Springfield Illionois. pp 372.

[16] Mukherjee PK (2002). Quality Control of Herbal Drugs. Business Horizons, New Delhi pp 131.

[17] Roychoudhury R, Saha C (1995). Contraception. In: Principles of Pharmacology. Munson PL, editor. New York: Chapman Hall pp 843-850.

[18] Singh HP, Singh CK, Singh RR (1998). Effect of potash alum (aluminium potassium sulphate) on human semen and sperm.Indian J Physiol Pharmacol 42:311-14.

[19] Yuan J, Yang D, Liang Y, Gao W, Ren Z, Zeng W, Wang B, Han J, Guo D (2012). Alkaloids from Areca (Betel) nuts and their effects on human sperm motility. Journal of food science 77; T70-T78 (9).

[20] Ejebe DE, Ojieh A E, Ovuakporaye SI, Odion-Obomhense HK, Adegor EC, Amadi CN, Nwadito C, Emudainohwo JOT, Ozoko TC (2008). Effects of anti-malarial alkaloids on the sperm properties and blood levels of reproductive hormones of adult men 7 (19); 3395-3400.

[21] Sur TK, Pandit S, Pramanik T (1999). Antispermatogenic activity of leaves of *Aegle marmelos* Corr. in albino rats: A preliminary report. Biomedicine19:199-202.

[22] Karou D, Savadogo A, Canini A, Yameogo S, Montesano C, Simpore J, Colizzi V, Traore AS (2006). Antibacterial activity of Alkaloids from *Sida acuta*. African Journal of Biotechnology 5(2): 195-200.

[23] Chen CR, Chao LH, Pan MH, Liao YW, Chang CI (2007). Tocopherols and Triterpenoids from *Sida acuta*. Journal of the Chinese Chemical Society 54: 41-45.

[24] Saraswathy A, Susan T, Gnana RR, Govindarajan S, Kundu AB (1998). Chemical investigation of *Sida acuta* Burm. Bull. Med. Eth. Bot. Res 19:176-180.

[25] Raja S, Sudhakar D (2000). Chemistry and pharmacological studied on *Sida* species- A review. Proc. Int. Cong. On Ayurveda-2000, Chennai, Tamilnadu, India 207-208.

[26] Ahmad S, Jamal Y, Mannan A (2011). Review of Some Medicinal Plants with Anti-fertility Activities. Unani Res. 1 (2): 24-28.

[27] Cao JH, Qi YP (1993). Studies on the chemical constituents of the herb Huanghuaren (*Sida acuta* Burm.f.). Zhongguo Zhong Za Zhi 18:681-682, 703.

[28] Gupta OP, Sharma ML, Ghatak BJ, Atal CK (1977). Potent uterine activity of alkaloid vasicine. Indian Journal of Medical Research 66: 865-871.

[29] Hiremath SP, Hanumantha Rao S (1990). Antifertility efficacy of the plant *Striga lutea* (Scrophulariaceae) on rats. Contraception, 42:466-477.

[30] Kumar D, Kumar A, Prakash O (2012). Potential antifertility agents from plants: A comprehensive review.Ethnopharmacology 140 (1):1-32.

[31] Wang JY, Ruan JL (1996). Triterpenes of *Marsdenia globifera* (for antifertility). Chin. Pharma. J. Zhongguo. Yaoxue. Zazhi 31, 266-269.

[32] Badami S, Aneesh R., Sankar S, Sathishkumar MN, Suresh B,Rajan S (2003). Antifertility activity of *Derris brevipes* variety coriacea. Journal of Ethnopharmacology 84: 99-104.

[33] Gebrie E, Makonnen E, Debella A, Zerihun L (2005). Phytochemical screening and pharmacological evaluations for the antifertility effect of the methanolic root extract of *Rumex steudelii.* Journal of Ethnopharmacology 96: 139–143.

[34] Brain KR, Ross RG (1977). An Introduction to Phytopharmacy: Steroid Hormones. Pitman Medical Publishing Co. Ltd., 42 Camden Road, Tunbridge Wells, Kent TN1 2QD: pp 150–158.

[35] Kumar P, Laloraya M, Laloraya MM (1989). The effect of some of the polyphenolic compounds on sperm motility *in vitro*: a structure activity relationship. Contraception 39: 531–539.

[36] Ghosal S, Srivastava AK, Srivastava RS, Chattopadhyay S, Maitra M (1981). Justicisaponin-I, a

new triterpenoid saponin from Justicia simplex. Planta Medica 42, 279-283.

CHAPTER FIVE

STUDY OF ELECTRONIC PROPERTIES OF TITANIUM DIOXIDE AND BARIUM TITANATE

Rashmi Rekha Mohanta*

Department of Chemistry, College of Engineering & Technology (A constituent college of Biju Patnaik University of Technology, Odisha), Bhubaneswar, India

ABSTRACT

The objective of the work is to study the electronic structure of high dielectric oxides of Ti based compounds like TiO_2, $BaTiO_3$ etc. deposited on Si(111) surface. When oxides are deposited on Si there may be possibility of interactions like metal-oxygen, metal-silicon, silicon-oxygen and metal-silicon-oxygen. All these interactions influence the physical and chemical properties that are important for technological applications. Therefore, investigation of these interaction strength is a major concern for device fabrication. Bulk $BaTiO_3$ was prepared by solid state reaction technique. TiO_2 and $BaTiO_3$ thin films on Si(111) surface were deposited by pulsed laser deposition technique. Both bulk and thin film samples have been investigated using X-ray Diffraction (XRD), Raman spectroscopy, X-ray Photoelectron spectroscopy (XPS) and Atomic Force Microscopy (AFM) techniques. Both XRD and Raman data suggest that the deposited TiO_2 is in anatase phase.

The binding energy peak position of Ti 2p also supports the formation of anatase phase. The AFM topography of as deposited film indicates the formation of non uniform TiO_2 growth with the formation of voids on Si(111) substrate which disappeared after sputtering with argon ion beam due to surface erosion. The Ti 2p core level of sputtered TiO_2 exhibits the formation of Ti_2O_3, TiO and pure Ti on the surface. After sputtering the high binding energy shoulder of O 1s peak becomes sharp. Ti LMM Auger peaks become broader but no shift in kinetic energy is observed. The XRD pattern of $BaTiO_3$ was indexed assuming a cubic structure with lattice parameter of 3.992A°.

The binding energy positions of Ba 3d and Ti 2p core levels observed in XPS studies indicated that Ba exist in 2+ valence state while Ti exists in 4+ valence state respectively. XPS results indicated a shift of the core levels of $BaTiO_3$/Si(111) to high binding energy by about 0.5 eV compared to the core levels of bulk $BaTiO_3$. Also the core level of the thin film are sharper compared to the core levels of bulk sample indicating more oxygen vacancies in bulk $BaTiO_3$ compared to thin film $BaTiO_3$. The AFM topography of $BaTiO_3$ on Si(111) surface indicates the formation of nano particles.

Keywords: Pulsed laser deposition, Thin films, Ion beam irradiation, X-ray Photoelectron Spectroscopy, Atomic Force Microscopy.

5.1 INTRODUCTION

Dielectrics are materials which do not conduct electricity but can sustain an electric field. All the dielectric materials are polarizable electric insulators. The insulators resist the electric current when electric potential difference is applied whereas a dielectric material gets polarised and stores electric energy. This is because, when a material is subjected to electric field, the response of the material depends mainly on the type of chemical bonding in the material. If the material possesses free electron as in metals or free charge carriers (electrons and holes) as in semiconductors, it conducts electricity because the charges are free to move through the materials. On the other hand in dielectric materials bound charges get slightly displaced when subjected to electric field. The cumulative effect of the microscopic displacement results in a net polarization due to the setting up of induced dipole moments or due to the rotation of permanent electric dipoles which are already present in the material. The molecules of a dielectric materials are either polar or non-polar. The polar molecules are those in which center of positive nuclei and negative electrons are separated by a small distance giving rise to permanent dipole moment. Unsymmetrical molecules such as H_2O, N_2O etc. are polar molecules. The non-polar molecules are those in which center of positive nuclei coincides with the centre of negative electrons. Examples of non-polar molecules are H_2, N_2, O_2 etc. Most of the electrical devices employ dielectric materials in some form or another. Current carrying wires and cables are always coated with dielectric material. Dielectric materials are used in the fabrication of actuators, transducers, rectifiers, amplifiers etc. Capacitors are filled with dielectric materials for storing electrical charges. Dielectric materials can also be used in liquid state.

Most of the dielectric materials are in the form of oxides. Dielectric oxides have a wide range of electrical properties[1]. Several dielectric oxides exhibit ferroelectric, piezoelectric and pyrroelectric properties. The range of dielectric constant goes from 2 to 150 for oxides. SiO_2, HfO_2, ZrO_2, TiO_2, $BaTiO_3$, $SrTiO_3$ etc are some examples of dielectric oxides. The dielectric constant of SiO_2 is 3.9. Oxides which have dielectric constant less than 3.9 are called low k-oxides and those oxides which possess dielectric constant more than 3.9 are called high-k oxides.

5.2 HIGH DIELECTRIC OXIDES

Most of the electronic devices such as Dynamic Random Access Memory (DRAM), Complementary metal oxide semiconductor (CMOS), Field effect transistor(FET) use SiO_2 as gate dielectrics in their capacitor structures. Dielectric layers play a crucial role in metal-oxide-semiconductor (MOS) devices which are used in the development of integrated circuits. Capacitor stacks used in integrated circuits occupy large areas of the circuit. The capacitance C of a parallel plate capacitor with dielectric layer completely filling the space between them is given by,

$$C = \varepsilon_0 kA/t$$

where A, the area of the capacitor plate, t the plate separation and ε_0, the permittivity of free space.

The area of these capacitors can be reduced at the cost of the thickness of the dielectric layer between the plates without compromising on capacitance. This means that to reduce the area of the capacitor by some factor the thickness need to be reduced by the same factor to retain the same capacitance. The conventional semiconductor technology uses SiO_2 whose dielectric constant is 3.9 as a gate dielectric in device fabrication. Scaling down the capacitor structures used in the devices requires SiO_2 layer of thickness 10A° or less. SiO_2 film of thickness less than 10A° causes leakage current to increase drastically [2]. The adverse increase of leakage current at smaller thicknesses is due to quantum mechanical tunneling of electrons through the silicon dioxide layer. The leakage current can exceed 1 A/cm^2 at 1 V leading to high power dissipation and malfunctioning of devices [3,4]. Moreover, it is challenging to produce and measure accurately such a small thickness films. Therefore the reduction of SiO_2 thickness is not a suitable solution for miniaturization of devices. Therefore, to fulfill the ambition of miniaturization of integrated circuits, higher dielectric (high-k) materials are required to replace SiO_2. Several high-k oxides such as TiO_2, $BaTiO_3$, $SrTiO_3$, ZrO_2, and Al_2O_3 to name a few were suggested as possible materials. As per Moore's Law the performance of CMOS devices improved over a period of forty years. The number of devices used in an integrated circuit exponentially increases, doubling in 2 or 3 year periods. Currently, miniaturization of electronic devices are in demand and therefore the minimum size of a transistor decreased exponentially with the passage of years. It is useful to express the thickness of new dielectric oxide in terms of its equivalent SiO_2 thickness or "Equivalent oxide thickness" (EOT) as EOT = $t_{ox} = (3.9/k)t_{hk}$ where, 3.9 is the dielectric constant of silicon dioxide. The objective is to develop high dielectric oxides with lower values of Equiv-alent oxide thickness (EOT).

Scaling down DRAM devices can also be achieved using high dielectric materials[5]. In DRAM, information is stored in the capacitor structure in the form of charge. These DRAMs need be refreshed periodically and the capacitor should be capable of storing charge for this period. This can be accomplished when the leakage current through the capacitor is $10^{-7} A/cm^2$. Silicon oxy-nitride is presently used as the capacitance dielectric, which need be replaced by a high k oxide to continue the scaling. The effective gate capacitance (C) is a series of combinations of three capacitors such as the oxide capacitor, the depletion capacitor of the gate electrode and the capacitor due to the carriers in the Si channel [4] and is give by

$$1/C=1/C_{ox} +1/C_{gate} +1/C_{Si} \quad (1.1)$$

As Capacitance varies with the thickness, 'effective capacitance thickness' of SiO_2 can be written as

$$ECT = EOT + t_{gate} + t_{Si} \quad (1.2)$$

Electronic industry uses Si for most of the integrated circuit applications. SiO_2 being a stable dielectric on Si is used as a dielectric in integrated capacitors. Silicon dioxide has the advantage that it grows easily as a native oxide by thermal oxidation whereas other semiconductors like Ge, GaAs, etc. has poor native oxide formation. Silicon dioxide is amorphous and has very few electronic defects, forms an excellent interface with silicon. As the miniaturization demands high dielectric oxides in place of SiO_2, we need to find a high dielectric which is stable on Si. Silicon is a semiconductor whereas silicon dioxide is an excellent insulator. The high dielectric should possess the following characteristics.

(a) It must have a high dielectric constant (k) and it should be thermodynamically stable with Si surface.

(b) It must be stable up to a processing temperature of 1000^0C for 5s.

(c) It should work like an insulator with band offsets to Si over 1 eV to reduce carrier injection in to its bands.

(d) It should form a good interface with silicon.

(a) High k value and thermodynamic stability

The first requirement implies that the choice of oxide should have 'k' value over 10 and preferably 25-30. The band offset condition requires a reasonably large band gap. It has been observed that k of the material candidate varies inversely with the band gap [6]. There are some oxides with very high 'k' such as $BaTiO_3$ (several hundreds of k value), but these possess too small band gap. However, a high k is undesirable in CMOS design as large fringing fields occur at source and drain electrodes [7]. The second requirement is that oxide should not chemically react with Si to form SiO_2 or a silicide according to the reaction.

$$MO_2 + Si \rightarrow M + SiO_2 \quad (1.3)$$

Thus, the interfacial SiO_2 would increase the EOT and narrow the effect of using the new oxide. The oxide with a higher heat of formation than SiO_2 is desirable for the purpose.

(b) Kinetic stability

High k oxide must be stable on Si surface during the processing where temperature goes to 1000^0C for few seconds. As an example, if we choose an amorphous oxide, it should remain amorphous when annealed up to 1000^0C for 5s.

(c) Band offset

The high dielectric oxide must have a band offset of about 1 eV with Si in order to inhibit conduction by the schottky emission of carriers into the oxide bands[6]. SiO_2 has high barriers for electrons and holes due to its wide band gap of 9 eV.

Materials with band gap less than 5 eV are not suitable for this purpose due to unsymmetrical alignment of both the barriers.

(d) Quality of the Interface

The oxide is in direct contact with the silicon surface. The carriers induced by the gate are introduced within a few A° of the silicon- oxide interface. Hence, the resulting interface must be of good and of defect free. Oxide grain boundaries are associated with extra defects. Thus, in order to get an interface with good quality one can use either a crystalline oxide grown epitaxially on the Si or an amorphous oxide.

The formation of inter-facial silicon dioxide layer is almost inveitable at the interface of the Si and high dielectric oxide. Such undesirable oxide layer with low dielectric constant leads to decrease in the capacitance. The overall equivalent oxide thickness of SiO_2 (layer-1) and high dielectric oxide layer (layer-2) is given by the formula as:

$$1/C=1/C_1 +1/C_2 \qquad (1.4)$$

which becomes

$$EOT = t_{SiO2} + EOT_{hik} \qquad (1.5)$$

Hence, an extra silicon dioxide layer is unwanted as it contributes to the overall EOT. In practice, the k value of SiO_2 is 3.9 and thus a SiO_2 layer can use up the equivalent oxide thickness and causes problems to down scaling the electronic devices. Some times interfacial oxide may have different composition than SiO_2 as Si can form other lower oxides. Interface can also be a combination of few silicon oxides along with SiO_2. Inter-facial oxide may not have the same quality as

SiO_2 prepared by thermal oxidation of Si [8]. Copel [9] studied the composition of inter-facial oxides under HfO_2 and found that some times SiO_2 appears to have higher k values than thermal oxide.

It may be useful, if we can control the thickness of the interfacial SiO_2 layer or remove it completely. There may be two ways. In first way, Si and SiO_2 react to form volatile SiO within the temperature 900 -1000^{0}C. The initial surface should be annealed to silicon oxide (SiO) [10]. The desorption of SiO occurs from a buried layer through a high dielectric oxide film. Another way is to react the metal like Hf with the SiO_2 to extract oxygen from SiO_2.

5.3 FORMULATION OF THE PROBLEM

Study of high dielectric oxides on Si surface is very important for technological applications which extensively use capacitor structures. As discussed before, to be a suitable candidate for high k applications, material should satisfy several conditions mentioned. When a high k oxide is deposited on Si surface as a thin film, how the properties of the oxide are modified as compared to the bulk oxide need be studied in detail. Many of the high k oxides have been thoroughly studied in their bulk form but they are not well explored in thin film form. Another important problem to be studied is the interaction of high k oxide with the substrate Si. The interface formation is more complicated due to the complexities involved in the chemistry and morphology of the interfaces. Most of the metal oxides react with Si and form SiO_2 or some sub oxide of Si at the interface. The thickness of the interfacial oxide depends on the roughness of the Si substrate and also on the chemical nature of the oxide. If the metal atoms in the dielectric are more electronegative compared to Si, it is expected that dielectric will not loose oxygen to Si surface. In such cases formation of Si oxides is not very favorable. There will be a competition between metal in the dielectric and substrate Si to gain oxygen to become oxides. Only oxygen is available in the dielectrics, so Si tries to gain oxygen. If the metal in the dielectric is less electronegative, it will give oxygen to Si, which leads to the formation of silicon oxides. In case if the electro negativity of the metal is more compared to that of Si, oxygen will not react with Si, but some metal silicide may form due to the diffusion of Si into the dielectric. The chemistry of interfaces depends not only on the chemical nature of deposit and substrate but also on geometry or the roughness of interface. Both chemistry and geometry of interfaces may vary from oxide to oxide. It is of utmost importance that the detailed study of surface and interface properties of high dielectric oxides on Si surface both for industrial applications and also for basic chemistry research point of view as well.

Chemistry of Ti is very interesting. It can react with different elements such as O, C, N, Si to form oxides, carbides, nitrides, silicides etc. Such multiple interactions are possible for Ti as it exhibits all possible valence states, starting from +1 to +4. It may also possess fractional valence states at surfaces and interfaces. Ti is extensively used in vacuum technology due to its unique property of sublimation where Ti metal is directly vaporized without going through liquid phase. This property is utilized in vacuum pumping, where sublimated Ti reacts with residual gases in the vacuum chamber and forms compounds which get deposited on the walls of the chamber. Chemistry of Ti is still not well explored from spectroscopic point of view.

In this work, two Ti based oxides, TiO_2 and $BaTiO_3$ have been investigated by depositing on Si surface. Chemical properties and technological importance of these oxides are briefly discussed below.

5.3.1 Titanium dioxide(TiO_2)

Titanium dioxide (TiO_2) is an interesting transition metal oxide possessing unique chemical and physical properties. It received enormous attention in modern technology as it is a wide band gap semiconductor and can be made into different forms like nano particles [11], nano-surfaces and bulk [12]. TiO_2 surfaces play a principal role in dye-sensitized solar cells as well as photo catalytic and photovoltaic applications [13], heterogeneous catalysis[14], waste water treatment[12]. Titania is preferred over other materials because of its abundance, low cost, photochemical stability, inertness and non-toxicity [15]. The unit cell and crystal structures of TiO_2 are shown in figure 1.1.

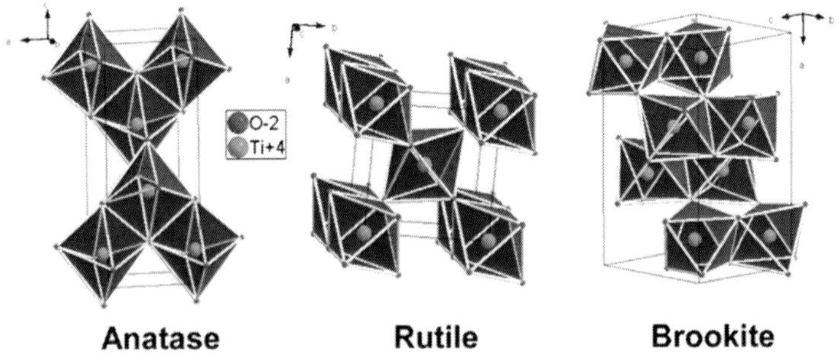

Figure 1.1: Different phases of TiO_2

Interestingly, TiO_2 can exist in three different stable structural forms called anatase, rutile and brookite as shown in figure 1.1. Both anatase and rutile are of tetragonal structures whereas brookite is orthorhombic [16]. The difference in their structures can be understood in terms of octahedral (TiO_6) distortion and connectivity of the octahedral chains [12]. The stability of the different TiO_2 crystal structures were discussed in detail by Navrotsky et al. [14]. However the following stability order has been found: rutile > brookite >anatase [17]. Titanium dioxide in anatase phase is suitable for making dielectric films in DRAMs (Dynamic Random Access Memory) like devices in electronic industry while rutile TiO_2 thin film exhibits a k value 100-170 [18] and are not suitable to the metal-insulator-metal(MIM) capacitors of DRAMs. The anatase TiO_2 film can be converted to the rutile phase by post annealing at temperature above 700^0C. Heterogeneous catalysis uses TiO_2 as a photocatalyst in solar cells for the production of hydrogen and electric energy. It is also used as a gas sensor, white pigment in paints and cosmetics, optical coating, corrosion-protective coatings, in ceramics and in electric devices such as varistors. It plays important role in the bio compatibility of bone implants, earth sciences. TiO_2 is an excellent gate dielectric for the new generation of Metal Oxide Semiconductor Field Effect Transistors (MOSFETs) and as a spacer material in magnetic spin-valve systems. Titanium dioxide finds applications in nanostructured form in Lithium based batteries and electrochromic devices.

Barium Titanate ($BaTiO_3$)

$BaTiO_3$ is one of the oldest ferroelectric known. $BaTiO_3$ can exist in cubic, tetragonal and hexagonal phases. Tetragonal phase becomes ferroelectric. $BaTiO_3$ is very stable in cubic structure at room temperature. Its stability on Si surface need be understood to know the feasibility of the material for electronic applications. The electronic properties of these materials are studied by electron spectroscopy.

5.4 ELECTRON SPECTROSCOPY

Most of the chemical and physical properties of materials depend on the electronic energy level spectrum of the constituent atoms. Direct measurement of electronic energy levels was not possible before 1960. Some techniques like X-ray absorption and emission used to measure the electronic energies in indirect ways. In the year 1960 Kai Siegbahn and his co-workers, used the basic photoelectric effect, to obtain the binding energies of electrons, in solids which fetched Kai Siegbahn the prestigious Nobel Prize in 1981[19]. Kai Siegbahn developed high resolution electron analyser using magnetic fields in Uppsala University, Sweden. In those days development of both high resolution electron analysers and ultra high vacuum system are novel and challenging. As the technique provides information on electron energy levels which can be used for chemical analysis of materials, it was named as Electron Spectroscopy for chemical Analysis (ESCA)[20]. This technique uses X-ray as the primary excitation

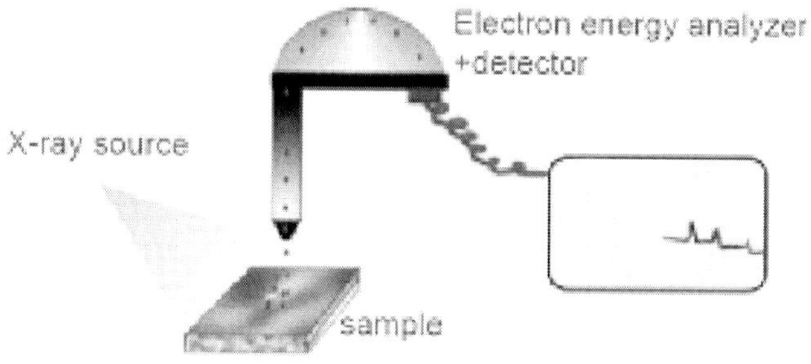

Figure 1.2: Basic XPS setup

source, and is also called as X-ray photoelectron spectroscopy(XPS) which is shown in figure 1.2. Using X-rays, deep core levels upto a binding energy of about 1500 eV can be probed. Though valence bands signals are also available in XPS, the intensity is very low due to weak photoelectron cross sections of valence band electrons in the X-ray region. The valence bands are usually studied by low energy ultraviolet radiation of energy between 20-40 eV range. At these low energies, line width of the radiation is very small giving rise to a better resolution. When ultraviolet radiation is used, the spectroscopy is called Ultraviolet Photoelectron Spectroscopy (UPS). There is a possibility for the emission of a second electron, called a Auger electron in the process of core hole decay in XPS. The Auger electron emission is more prominent when the X-ray beam is replaced with an electron beam. The spectroscopy involving the studies of Auger electrons is called Auger Electron Spectroscopy (AES). Electron Spectroscopy is a general phrase used for XPS, UPS and AES. When an atom or a molecule (M) absorbs a photon, it gets excited to a meta stable state, M. In the deexcitation process M emits an electron and becomes positive ion (M^+), as described in the equation below.

$M + h\nu \rightarrow M^* \rightarrow M^+ + e^-$

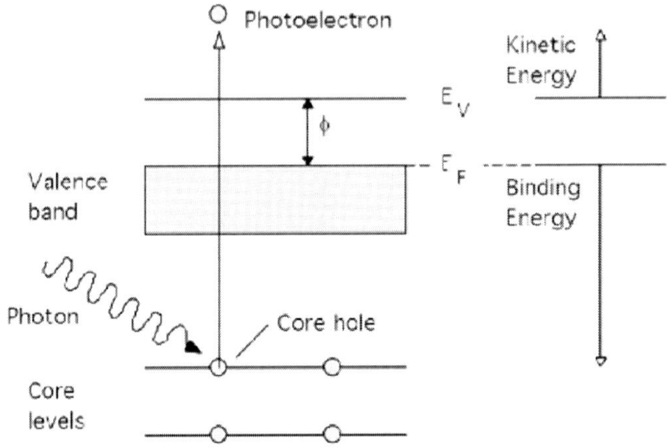

Figure 1.3: Photoemission Process

The above process is very similar to the explosion of a bomb when heat is supplied. In the above M^* explodes into two fragments M^+ and e^-. In this process both momentum and energy are conserved. Since M^* is at rest, initial momentum is zero. After explosion both M^+ and e^- have the same magnitude for the momentum in opposite directions. If the magnitude of momentum is p, then the kinetic energy carried away by M^+ and e^- respectively are $p^2/2M$ and $p^2/2m$ where M and m are the masses of the ion and electron respectively. Mass of the ion, M^+ is same as the mass of the molecule, M. As the mass of the ion is very large compared to the mass of the electron (M >> m), the kinetic energy of the ion is negligible compared to the kinetic energy of the electron. Even in the case smallest possible ion i.e. H^+, the ion mass is 2000 times the electron mass. For all practical purposes the kinetic energy of the ion can be neglected. The photon energy hν is used entirely for dislodging the electron from its bound state and supplying kinetic energy (E_K) to it. The energy required to dislodge the electron is called its binding energy (E_B). Thus, $h\nu = E_k + E_B$. In case of solids, electrons require an additional energy called work function (ϕ) to come out of the surface. Therefore the kinetic energy of the emitted electron for solids should be written as

$$E_k = h\nu - E_B - \phi.$$

Ultra high vacuum (pressure less than 10^{-8}Torr) is essential for conducting electron spectroscopy experiment. Such a vacuum could not be achieved prior to 1960 and thus blocked the development of electron spectroscopy technique, though the basic principle of photo emission was understood by 1905. The principle of photoemission process is shown in figure 1.3. The stringent condition on vacuum is to maintain clean sample surface during the conduct of the experiment which may require about an hour. Major concern in the spectroscopy is the time required for the formation of one monolayer contamination on the surface. This time needs to be increased upto an hour or so which is possible only when the vacuum is better than 10^{-8}Torr.

5.4.1 Mean free path and surface sensitivity

The photoelectrons which are detected in photoemission experiment originate only from the top most layers of a solid i.e the first few monolayers of the surface of a solid. X-rays can penetrate through solids upto few microns and ionize the atoms of the solid by dislodging the electrons having binding energies smaller than that of incident X-ray. Photoelectrons emitted at deeper layers, loose energy while reaching the surface. Whereas the electrons originated in the near surface region reach the surface without loosing considerable energy.

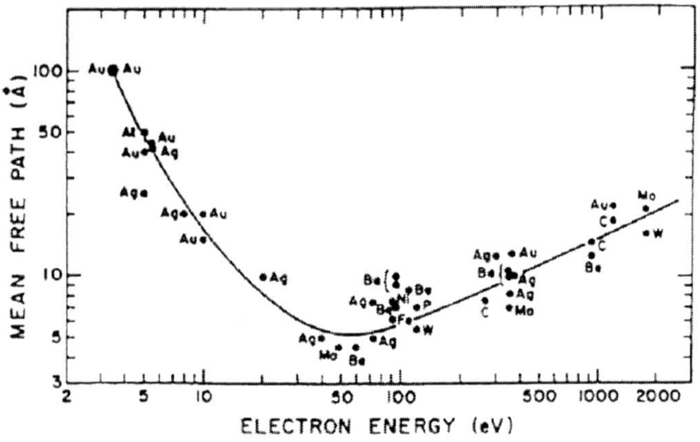

Figure 1.4: Energy dependence of mean free path

Those photoelectrons which do not loose energy while travelling to surface only represent the actual electron state of origin. The photoelectrons which loose energy in the solid have lost the memory of their original energy state. There is a characteristic length called mean free path (MFP) which depends both on the solid and the kinetic energy of the electron. Photoelectrons which are emitted in a depth less than IMFP will reach the surface without losing energy. Whereas the electrons emitted below a depth greater than MFP loose their energy, due to inelastic scattering. MFP has been measured for many solids at different kinetic energies and interestingly the data for various solids follow a universal curve shown in figure 1.4. Prominent mechanism of energy loss by electrons in a solid is due to electron-electron interactions. This mechanism dominates for electrons having energy greater than 50 eV. At low kinetic energies (< 50 eV) the primary mechanism of energy loss is electron-phonon interactions. Photoelectrons with high energy travelling in solids can excite other electrons to empty states, i.e it can excite a valence band electron or a core electron to empty state and loose its energy. This surface sensitivity is due to the strong interaction of electrons with matter. An electron moving inside a solid will have certain inelastic mean free path i.e a characteristic length that electron can move without suffering an energy loss.

The result of these processes is to make the mean free path of electron in a solid strongly dependent on its kinetic energy. At very low kinetic energies these processes are not excited as the energy is not sufficient, as a result its mean free path is long. On the other hand, at high kinetic energies the electron spends less time in passing through a given thickness of solid and thus there is less energy loss. Thus its mean free path in the solid is again quite long. However the mean free path passes through a minimum between these two regions. In photoelectron spectroscopy even though the penetration depth of the incident X-rays is typically of the order of microns, the electrons that are ejected from the solid will have originated only from the top few layers due to their mean free path.

Auger Electrons

A core hole is created by ejection of an electron from a core level by a photon (X-ray) in conventional photoemission. The core hole is filled by an electron transition from a higher energy level. The energy available in this transition may come out as radiation or another electron from one of the upper levels.

The secondary electron ejected in a non-radiative process is called auger electron. The process in which radiation comes out is called X-ray fluorescence (XRF). Auger electron emission and X-ray fluorescence will occur as a consequence of creation of core hole. The processes are shown in the figure 1.5.

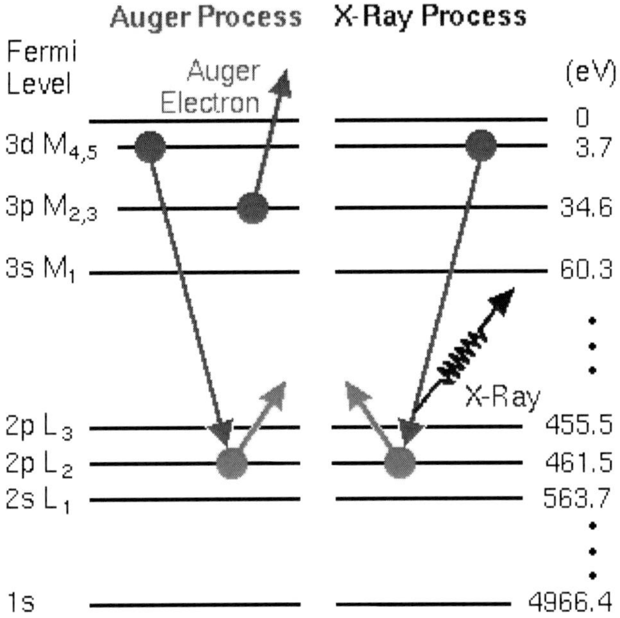

Figure 1.5: Auger electron and X-ray fluorescence processes

In XPS Auger electrons are ejected along with the photoelectrons. The Auger electrons are called X-ray excited Auger electrons and the corresponding spectroscopy is termed as X-ray excited Auger electron spectroscopy (XAES).The Auger electron emission and X-ray fluorescence are competing events. One of them occurs at the cost of others. The Auger electron and the fluorescence yields are depicted in the figure 1.6. Auger yield is more likely for core hole in elements with low atomic number(Z) almost exclusively for $Z < 15$ while XRF is more likely for deep core hole with high atomic number(Z) elements. The kinetic energy of Auger electrons are characteristicsof the atom of origin and act as finger print of the atoms. The designation associated with Auger transition is based on the shell nomenclature. As an example if the valence levels are involved in the Auger process, these are denoted by V. Auger electrons are denoted by X-ray notation as KLL, $KL_1L_{2,3}$, Figure 1.6: Auger electron and fluorescence yield $L_{2,3}M_{2,3}M_{4,5}$ or KVV.....etc. There are three levels involved in Auger process. If the binding energies of the levels involved are E_i, E_j, E_k and the original hole is created in the level i and the hole at i is filled by the electron in j and suddenly electron from level k comes out.

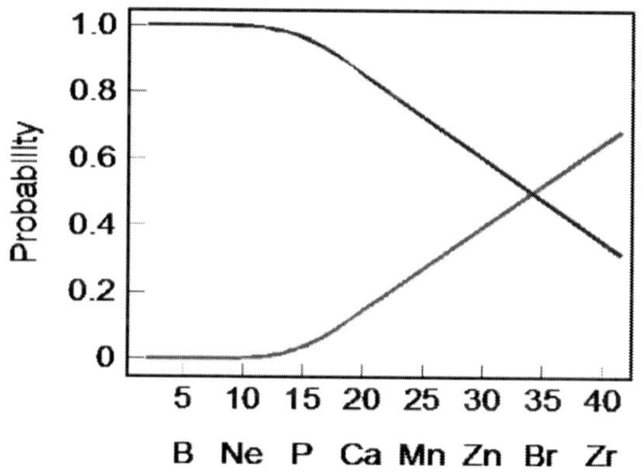

The kinetic energy of the Auger electron is given by the equation

$$E_{ijk} = E_i - E_j - E_k^*$$

E_k^* represents the Binding Energy of the level k with a hole in the level j. The Auger transition corresponding to the level is represented as ijk. For example, we observe KLL Auger transition whose energy is given by

$$E_k - E_{L1} - E_{L2,3}$$

2. METHODOLOGY

This section describes the methods used for sample preparation and techniques used for characterizing the prepared samples. Bulk and thin films of TiO_2, $BaTiO_3$ were prepared and were characterized using X-ray Diffraction(XRD), Raman Spectroscopy, Atomic Force Microscopy(AFM) and X-ray Photoelectron Spectroscopy(XPS) techniques. Bulk samples have been prepared by solid state reaction technique and thin film samples were prepared by Pulsed Laser deposition (PLD) technique.

2.1 Solid-state reaction technique

Solid state reaction technique involves the chemical reaction between two solids in the powder form, without using other solvents. This method is well suited for preparation of oxide powders. In this technique, reactants are in the form of powders and are mixed thoroughly using pestle and mortar to get a homogeneous mixture. Such a mixture is calcined at high temperatures of about 2/3 of the melting point of the reactant with lower melting point. Calcination need be done for few days as the reactants in the form of solid react only by diffusion which is very slow process. To ensure homogeneity, intermediate grindings are done during calcination. Calcination period, calcination temperature and number of grindings vary from sample to sample. In general, mixtures need be calcinated for a day or two with calcination temperatures usually above 900^0C. Samples prepared by this technique are free from impurities as no other solvents are used. In this process, only polycrystalline powdered samples can be produced. Quality of the samples can be improved by increasing the grain size which is done by pelleting the calcined powder and sintering the pellet for several hours at the calcination temperature. After sintering, hard pellet samples are produced which can be used for characterization. Amount of reactants should be taken in the ratio required to give the stochiometry of the compound. For example $BaTiO_3$ can be prepared by taking $BaCO_3$ and TiO_2 as initial reactants and the amounts required are calculated using the chemical reaction,

$$BaCO_3 + TiO_2 \rightarrow BaTiO_3 + CO_2$$

During the calcination process, it is assumed that Ba and Ti contents remain same, problem arises if one of the reactants is volatile (for example Pb). Volatile reactant is lost during calcination and the stochiometry of the final product will not be same as expected from the reaction. In such cases one should take little more amount for the volatile reactant. It is not easy to estimate, how much more need to be taken. Therefore, for compounds with volatile substances, enough care should be taken to maintain stochiometry. Samples used here are devoid of such volatile substances. $BaTiO_3$ bulk sample is prepared by this technique using high purity $BaCO_3$ and TiO_2 powders as starting chemicals. Binding solvents like poly vinyl alcohol was not used for making pellets as such materials may react with compound and give furious signals in photoelectron spectroscopy.

2.2 Pulsed Laser Deposition (PLD)

Pulsed Laser Deposition (PLD) is most used technique to deposit thin films of compounds. This technique uses a bulk target material of the compound. Bulk target is ablated using a laser beam. Laser ablation produces the cluster of particles of the target material. These ablated particles are deposited on a substrate of interest. PLD method produces thin films with the same stochiometry as that of target material. Thus PLD is widely used for preparation of complex oxide thin films where oxygen partial pressure in the chamber helps to maintain the stochiometry in thin film. This technique is found to be very suitable to prepare high temperature superconductors like Yttrium Barium Copper Oxide (YBCO) [21]. The substrate $BaTiO_3$ is prepared by PLD.

2.2.1 Basic principle

In Pulsed laser deposition technique, a high-energetic laser beam is focused on a target material which results in ablation of the material. A dense layer of vapor is formed in front of the target at the early stage of the laser pulse. The energy of absorption during the remainder of the laser pulse causes increase in temperature and pressure of this vapour resulting in partial ionization. Due to the increase in pressure, this layer expands from the target surface and forms "plasma plume". Also during this expansion, internal thermal and ionization energies are converted into the kinetic energy of the ablated particles. During expansion into low pressure back ground gas, attenuation of the kinetic energy occurs due to multiple collisions. Usually, for the deposition of complex oxides a back ground pressure of 1 to 50 pa is used. Thermalization occurs at these pressures at a penetration length comparable to the target to substrate distance. Numerous models like the drag "force" and the "shock wave" model [22] have been proposed to describe this attenuation in low and high back ground pressure respectively.

The parameters, mass and pressure determine the interaction with the ablated particles and also the kinetic energy of the particles reaching the substrate. As a result of which this kinetic energy can be varied from high initial energy in vacuum to low energies obtained from thermalization at sufficient large ambient pressure. This is one of the unique feature of PLD which can be used to modify thin film growth. Here, the controllable parameters of the energetic particles at the film surface are diffusivity [23,24] and the probability of both absorption and desorption [25].

High deposition rate is the another unique feature of PLD. Usually, deposition rates range from 10^{-2} to 10^{-1} nm/pulse with deposition pulse duration in the order of several μs to 500 μs [26, 27] in Pulsed Laser Deposition of oxides. As a consequence the deposition rate can be as high as 10^{2} to 10^{5} nm/sec [23, 24]. The order of magnitude of this value is higher as compared to other physical vapor deposition techniques, like molecular beam epitaxy and sputter deposition in which deposition rates range from 10^{-2} to 10^{-1} nm/sec.

In the experiments a KrF excimer laser (λ = 48nm) with maximum pulse rate of 50Hz is used. The pulse duration is nearly 20ns with maximum pulse energy 650mJ. To select the homogeneous part of the laser beam a mask is used. The mask is focused at an angle of 45^0 on the target by means of a focusing lens. The substrate holder, target holders and heater are mounted on a computer controlled XYZ- rotation stage and base pressure of the vacuum is maintained at 10^{-5}pa. Oxygen gas as well as inert gases can be used as deposition gases. The deposition pressure is controlled by the effective pump speed and the speed is adjusted through a variable restriction between the deposition chamber and turbo-molecular drag pump.

2.3 Deposition of TiO_2 and $BaTiO_3$

In PLD of complex oxides, stochiometric transfer of particles from multi-component targets to substrate is much essential. However, this transfer is affected by the target properties like homogeneity and density. During ablation the density of target materials should be as high as possible to avoid removal of big clusters from the target. Single crystals are preferably used as target material because of the high density and homogeneous distribution of components. However single crystals are not easily obtained and therefore sintered pallets with highest possible density are used.

In our experiments, we used polycrystalline pellets of TiO_2 and $BaTiO_3$. In this study, we have used commercially available high purity (99.99%) TiO_2 powder. $BaTiO_3$ targets is prepared by solid state reaction technique. Finally the pallets were made into thin films on Si substrate by pulsed laser deposition technique. Both the oxygen deposition pressure and substrate temperature are selected to maintain the stochiometric growth of the films.

2.4 X-ray Photoelectron Spectroscopy (XPS) measurements

In this work two high dielectric oxide systems TiO_2 and $BaTiO_3$ have been extensively studied using XPS technique. XPS measurements were carried out at Institute of Physics, Bhubaneswar using VG ESCA machine equipped with the hemispherical electron analyzer, twin anode X-ray source, and Argon ion gun. This machine has a load lock system to introduce samples without disturbing the vacuum in the analysis chamber. The load lock system is pumped by a turbo molecular pump; and is maintained at a base pressure of 5.0 x 10^{-8} Torr. The analysis chamber is pumped by Ion pump and is maintained at a base pressure of 3.0 x 10^{-10} Torr.

Samples have been loaded on to the sample holder using double sided conducting tape. Conducting tape helps in connect the sample to electrically ground and thus neutralizes the sample during photoemission. The sample holders have been transferred to the load lock by venting the load lock chamber with N_2 gas. The load lock chamber has been pumped for overnight to obtain a vacuum of low 10^{-7} Torr. Then the sample holder has been transferred to analysis chamber which is at a base pressure of low 10^{-10} mbar.

X-ray source is switched on with 10 mA filament current and 15 KV anode potential. Sample is kept at the focus of the analyser and X-ray source. Energy ranges for the scans have been set in the soft ware of the computer connected to the system. Fixed Analyser Transmission (FAT) mode has been used for all the data used in this thesis. In FAT mode instrumental energy broadening is same at all energies.

For each sample a survey scan is recorded from 1200 eV - 0 eV using 20eV pass energy and 0.6 eV step sizes. Individual high resolution scan were recorded with 0.05 eV step and 20 eV pass energy. Dwell time used in all the scans is 0.1s.

XPS data has been analyzed using XPSPEAK software [28]. Core level spectra have been fitted with Gaussian-Lorentzian sum function, given by

$$GL(x, p, w, h, m) = h\{(1 - m)e^{-Q\ln 2} + m/(1 + Q)\} \quad (2.1)$$

where $Q = [\ 2(x-p)/w\ ]^2$; p, the peak position; w, the FWHM of the peak; h, the integrated area of the peak; m, the GL mixing parameter (0 for 100% Gaussian and 1 for 100% Lorentzian);

2.5 Raman Spectroscopy

Raman Spectroscopy is versatile, convenient, non-destructive characterization tool to obtain information about the molecular or ionic structure and the chemical bonding between atoms. Since it probes lattice vibrations, which are sensitive to internal and external perturbations, it can be used to study the composition, phase, crystallinity, crystal orientation and in some cases doping in semiconductors. An important aspect of Raman Spectroscopy is its ability to determine the changes in the vibrational frequencies induced by the microstructure and by the strain in the lattice[29,30] Thus, it is a very useful tool for studying structural modification in dielectrics after ion irradiation[31, 32]. Micro-Raman spectroscopy, where the light of exciting laser is focused on the sample through a microscope, is especially very useful for microelectronic devices with small dimensions as it allows investigation with micron spatial resolution. Raman spectroscopy is inelastic light scattering technique used to diagnose the internal structure of molecules and crystals [29,30]. The light of a known frequency and polarization is scattered from a sample. The majority of the scattered photons have the same frequency as incident photons and is known as Raleigh scattering. The Raman scattered photons have frequencies shifted from the incident photons by energies of molecular vibrations. The magnitude of the shift is independent of the excitation frequency. This “Raman shift” is therefore intrinsic property of the sample.

Most of the Raman scattered photons are shifted to longer wavelengths known as stokes shift. But a small portion is shifted to shorter wavelengths called anti-stokes shift. In each case, the incident photon excites an electron to a higher energy level and then the electron deexcites back to a lower level, emitting a scattered photon. In Rayleigh scattering the electron decays is back to the same vibrational level from which it is started. Stokes Raman Scattering occurs when the final energy level is higher than the initial level, while anti-stokes Raman scattering occurs when the final energy level is lower than the starting level. Stokes scattering is much more common than anti-stokes scattering because at any given time an electron is most likely to be in its lowest energy state at room temperature. Only stokes Raman scattering is commonly used in spectroscopy.

2.6 Atomic Force Microscopy (AFM)

Atomic force microscopy (AFM) is an extensively used technique in the fields of fundamental research, technology and bio-medicine. It does not require the sample to be electrically conductive unlike scanning tunneling microscopy (STM). AFM can be carried out on any type of material like insulators, semiconductors and electrical conductors. As its probe is very small and sharp, AFM can be used to acquire high-resolution real space topographic images of sample surfaces. Hence it is a very effective technique to monitor/characterize the morphology and topological changes of ion implanted surfaces.

The atomic force microscopy (AFM) probes the surface of a sample with a sharp tip, a couple of microns long and often less than 100A° in diameter. The tip is located at the free end of a cantilever that is 100 to 200 micro meter long. Forces between the tip and the sample surface cause the cantilever to bend, or deflect. A detector measures the cantilever deflection as the tip is scanned over the sample, or the sample is scanned under the tip. The measured cantilever deflections allow a computer to generate a map of surface morphology.

Several forces typically contribute to the deflection of an AFM cantilever. The force most commonly associated with atomic force microscopy is an interatomic force called the vander Walls force.

There are mainly three scanning modes used in AFM. These are the following:

(i) Contact or repulsive mode:

Contact AFM mode is also known as repulsive mode, an AFM tip makes soft 'physical contact' with the sample. The tip is attached to the end of a cantilever with a low spring constant, lower than the effective spring constant holding the atoms of the sample together. As the scanner gently traces the tip across the sample (or the sample under the tip), the contact force causes the cantilever to bend to accommodate changes in topography. This mode is applied on hard crystalline surface with atomic or molecular resolution, but it is not suitable for biological samples where this force is still strong enough to cause their deformations.

(ii) Non-contact or attractive mode:

Non-contact AFM is one of several vibrating cantilever techniques in which an AFM cantilever is vibrated near the surface of a sample. The spacing between the tip and the sample is on the order of tens to hundreds angstroms. In this mode, the tip is at a distance from the sample surface where the attractive Vander Walls forces dominate. The major advantage of the non-contact mode AFM is its truly non-invansive character. This mode is suitable for biological samples or soft samples.

(iii) Tapping mode or Intermittent-contact atomic force microscopy:

In intermittent-contact atomic force microscopy, the vibrating cantilever tip is brought closer to the sample so that at the bottom of its travel it just barely hits, or 'taps', the sample. This mode is a compromise between the contact and non-contact modes. The cantilever is made to oscillate so the tip is very close to the sample for a brief time. The compromise between the two modes allows one to scan soft adsorbate on a substrate with better resolution than in the non-contact mode but with less interaction (and, consequently, less modification of the sample) between the tip and adsorbate as in the non-contact mode.

2.7 X-Ray Diffraction (XRD)

X-ray diffraction (XRD) technique is used to analyze the crystal structure of the materials. X-rays are electromagnetic radiation with photon energies in the range of 100 eV to 100 KeV. For diffraction applications, only short wavelength x-rays in the energy range of 1 KeV to 120 KeV are used. X-rays are ideally suitable for probing the atomic structures of materials because, the wavelengths (0.1 to 2°A) are comparable to the radii of the atoms and they are sufficiently energetic to penetrate most materials to provide information about the structure of the material. In powder samples, the crystalline domains are randomly oriented in 3-Dimensional space. Therefore, when the 2-Dimensional diffraction is recorded, concentric rings of scattering peaks corresponding to the various lattice spacings(d) in the crystal are observed. The positions and intensities of the peaks are used for identifying the underlying structure (or phase) of the material. In case of single crystalline sample, the crystalline domains are oriented, rather in single planes. X-ray diffraction uses the property of crystal lattices to diffract monochromatic X-ray. This involves the interference of waves scattered at the succssive planes, which is described by Bragg's equation:

$$n\lambda= 2d\sin\theta(n =1,2,3...)$$

where λ is the wavelength, d is the lattice plane separation and is half of the diffraction angle.

The diffractometer consists of a Goniometer, X-ray source, and a detector. CuK_{α} radiation ($1.5415A^{0}$) is used as the X-ray source and scintillation counter as the detector. The Goniometer consists of a fixed X-ray source, which requires that the detector and the sample tilt simultaneously to render the θ-2θ angular relationship between source and the detector. The system has a wide-ranging step and continuous scan capability with programmable step rotation.

3. DISCUSSION

Figure 3.1 shows 1000 nmX1000 nm two dimensional images of as deposited (figure 3.1(a)) and sputtered (figure 3.1(b)) samples. As deposited film exhibits non uniform growth and lot of voids or pits formed on the surface. Surface morphology of the as deposited sample indicates a discontinuous film growth on Si(111). A large number of voids of various sizes are observed on the surface which are randomly situated. After sputtering, voids are not so clear due to the erosion of the film caused by the ion beam. Distribution of void size has been determined for as deposited sample and the average size is found to be 83 nm as shown in the inset of figure 3.1(a). Such void size distribution could not be carried out on the sputtered surface as most voids disappeared due to surface erosion. TiO_2 films prepared at different temperatures exhibit different surface morphology and the films prepared at higher temperature exhibit a non uniform morphology [33]. In our case the substrate temperature during film growth is maintained at 500^0C and thus deposited clusters have mobility so that they migrate to different locations to form stable structures.

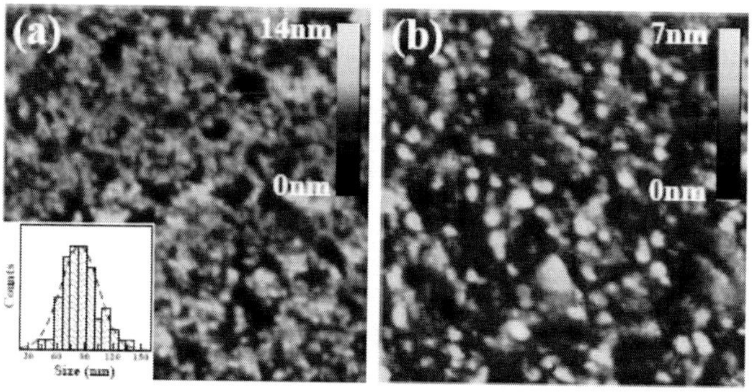

Figure 3.1: AFM topography (a) as deposited (1000 nmX1000 nm) (inset shows pit size distribution) and (b) sputtered (1000 nmX1000 nm) TiO_2/Si(111)

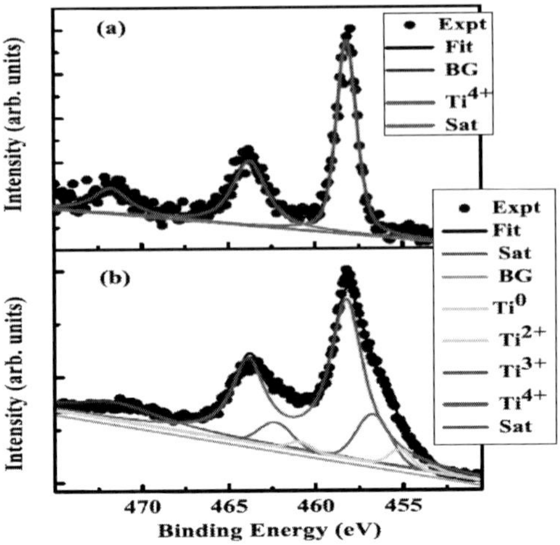

Figure 3.2: Ti 2p photoelectron line of (a) as deposited and (b) sputtered TiO_2/Si(111) recorded using AlK_α. BG in the figure legend indicates the linear background and Sat indicates satellite.

Ti 2p core level of as deposited and sputtered TiO2/Si(111) is shown in figure 3.2. These spectra have been numerically fitted using GL sum function defined by equation 2.1. The as deposited sample (Figure 3.2(a)) exhibits a sharp and intense spin orbit doublet of Ti 2p with $2p_{3/2}$ peak at 458.2eV and a spin orbit splitting (S.O.S) of 5.6eV which indicate the formation of stoichiometric TiO_2[34,35]. A sharp Ti $2p_{3/2}$ peak with FWHM of 1.5eV supports observed XRD and Raman for the formation of anatase TiO_2 without other phases of TiO_2. Ti 2p peak broadens when both rutile and anatase phases are existing in the sample as observed by Scanlon et.al [36]. The FWHM of Ti $2p_{1/2}$ is about 2.3eV which is quite large in comparison to $2p_{3/2}$ peak due to Coster Kronig transitions. A high binding energy shake up satellite is observed at 471.5eV. Various satellites of Ti 2p core level observed in TiO_2 and $SrTiO_3$ were explained in detail by Kim & Winograd [37]. According to them TiO_2 exhibits two satellites for Ti $2p_{3/2}$ core level at 13.0eV and 25.4eV binding energies from the main peak. The satellite we observed is at 13.3eV higher in binding energy (471.5eV) from the main peak (458.2eV). This satellite can be attributed to a shakeup satellite of $2pe_g \rightarrow 3de_g$ type monopole charge transfer [37]. Such satellite formation is observed only in Ti compounds with Ti^{4+} valence state. After sputtering, Ti 2p spectrum (Figure 3.2(b)) exhibits 4 doublets with each doublet having the S.O.S of 5.6eV. The $2p_{3/2}$ of the anatase TiO_2 (Ti^{4+}) occurs at 458.2eV [38, 39]. Three low binding energy shoulders to $2p_{3/2}$ main peaks at 456.7eV, 455.2eV and 453.4eV correspond to Ti_2O_3 (Ti^{3+}), TiO (Ti^{2+}) and pure Ti (Ti^0) respectively. Binding energy positions of various valence states of Ti agree with the values provided by M.C. Biesinger et.al [40] within the error bars given by them. Interestingly, the intensity of the shake up satellite at 471.6eV came down drastically after sputtering due to the reduction in concentration of TiO_2 (or Ti^{4+} valence state)on the surface. The ion beam sputtering preferentially removes oxygen compared to Ti due to higher sputtering rate for oxygen [41]. Such differential sputtering causes TiO_2 reduce into Ti_2O_3, TiO and pure Ti on the top surface of the sample.

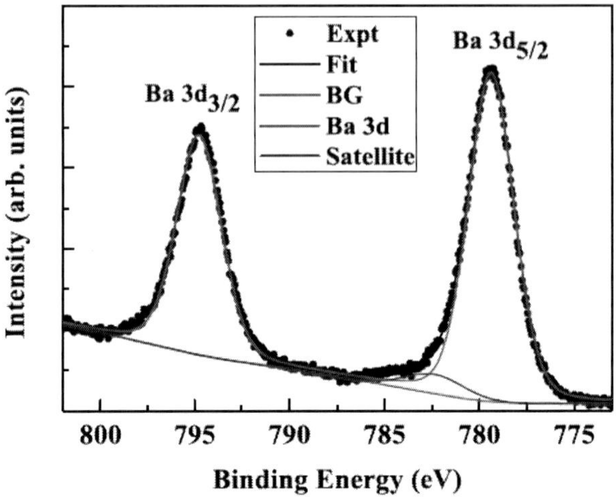

Figure3.3: Ba3d core level of bulk $BaTiO_3$ using AlK_α radiation.

Ba 3d core level of bulk $BaTiO_3$ recorded using AlK_α radiation is shown in figure 3.3. The Ba $3d_{5/2}$ peak occurs at 779.4eV. The spin orbit splitting (SOS) of the doublet is 15.3eV. The peak position and SOS values agree with the 2^+ charge state of Ba in $BaTiO_3$. A satellite peak of Ba $3d_{5/2}$ is observed at 784.9eV i.e 9.5eV lower in binding energy with respect to $3d_{5/2}$. Such satellites occur as the unmonochromatic $AlK_\alpha 3,4$ radiation due to ghost X-ray. The FWHM of the Ba $3d_{5/2}$ is about 2.7 eV with TS and TL values 0.15 and 40 respectively which indicate the asymmetry in the core levels

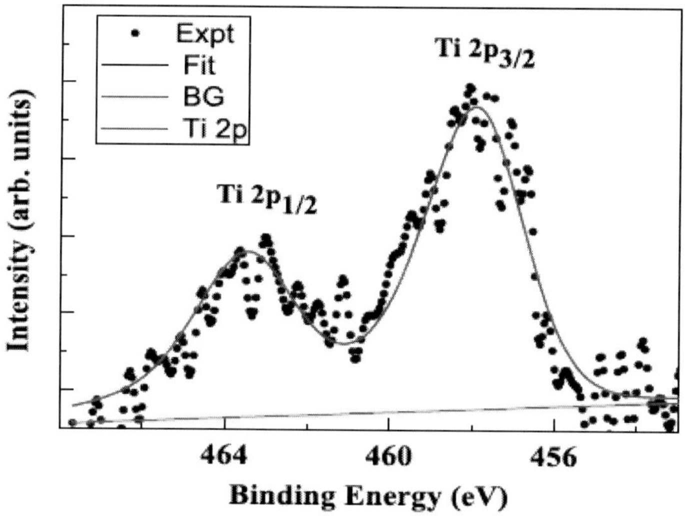

Figure 3.4: Ti 2p core level of bulk $BaTiO_3$ using AlK_α radiation. BG indicates the Shirley background.

Figure 3.4 shows the Ti 2p core level of bulk $BaTiO_3$. The $2p_{3/2}$ component of Ti 2p occurs at 457.9eV with a spin orbit splitting of 5.6eV which agrees with the corresponding value of 4^+ charge state of Ti. 4^+ valence state is stabilized for Ti in $BaTiO_3$ as it sits at the center of TiO_6 octahedra. TS and TL values obtained from numerical fitting with AGL sum function are 0.2 and 70 respectively.

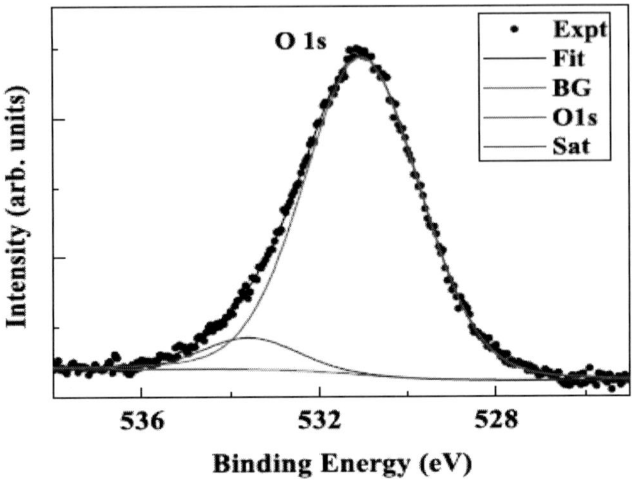

Figure 3.5: O 1s core level of bulk $BaTiO_3$ using AlK_α radiation. BG indicates the Shirley background

O 1s core level of bulk $BaTiO_3$ is shown in figure 3.5. The main peak at 531.5eV correspond to stoichiometric oxygen in the sample and the shoulder at 533.5eV occurs due to adsorbed hydroxyl species and also due to oxygen adsorbed in the grain boundaries. Observed XPS spectra of Ba 3d, Ti 2p and O 1s of $BaTiO_3$ agree well with the reported spectra[42-44].

The two dimensional 1000 nm×1000 nm AFM image of $BaTiO_3$ thin film on Si(111) is shown in figure 3.6. The image indicates the formation of islands due to the lattice mismatch between Si substrate and $BaTiO_3$. The average island size is about 20 nm.

Figure 3.6: AFM topography (1000 nm×1000 nm) of $BaTiO_3$ thin film on Si(111) surface

CONCLUSION

In conclusion two oxide systems, TiO_2 and $BaTiO_3$ thin films on Si(111) surface and $BaTiO_3$ bulk samples have been investigated using X-ray Photoelectron spectroscopy (XPS) and Atomic Force Microscopy (AFM) techniques. All the samples have been characterized structurally using X-ray Diffraction (XRD) and Raman spectroscopy. Bulk powders have been prepared by solid state reaction method which has advantages such as the low-cost raw materials and impurity free simple process. In this method, starting chemicals are thoroughly mixed and grounded to obtain a uniform fine mixture and calcinated at a temperature below the melting point of chemicals for several hours. The resulting powder was then made in to pellets and sintered for 12 hr at very high temperature (1250^0C). The thin films were deposited on Si(111) surface from these pellets by conventional PLD method which is considered as one of the best technique for stochiometric deposition of thin films. Prior to the deposition Si substrate was degassed by flashing at 900^0C in ultra high vacuum (UHV) to remove surface oxides. The deposition chamber was back filled with oxygen to a pressure of $1.0x\ 10^{-5}$ mbar during ablation to maintain oxygen stochiometry in the deposited film. The bulk as well as thin films were structurally characterized by X-ray Diffraction technique (XRD) and Raman spectroscopy. Electronic properties of these oxides were investigated by X-ray Photoelectron spectroscopy (XPS) using a VG ESCA machine equipped with argon ion source, twin anode X-ray source and a hemispherical analyzer. Ion beam sputtering was done using a 3 KeV argon ion beam in-situ in the preparation chamber maintained at a base vacuum of 2.0×10^{-8} mbar for TiO_2/Si(111). Argon ion current was fixed at 10 μA using an argon pressure of 5.0×10^{-6} mbar. Ti 2p, O 1s, wide range Valance band (VB) region including Ti 3s, Ti 3p, O 2s core levels and Ti LMM Auger spectra were recorded along with wide energy survey scans in Fixed Analyzer Transmission (FAT) mode using AlK_α radiation. The total resolution of the spectrometer is 0.8eV. AFM studies have been done on as deposited and sputtered samples. AFM (Nanoscope III, Veeco) measurements were done in tapping mode using SiN tip immediately after taking out the samples from XPS chamber. The XRD along with Raman indicated that the deposited TiO_2 is in anatase phase which is further supported by the binding energy peak position of Ti 2p in XPS. As deposited sample exhibit a sharp Ti 2p doublet with Ti $2p_{3/2}$ at 458.2 eV confirms 4^+ oxidation state of Ti. A high binding energy shake up satellite is observed at 13.3 eV higher in binding energy from the main peak attributed to $2pe_g \rightarrow 3de_g$ type monopole charge transfer transition. After sputtering with Argon ion beam, TiO_2 was reduced to Ti_2O_3, TiO and pure Ti metal on the surface. The sputtered surface contains Ti^{4+}, Ti^{3+}, Ti^{2+} and Ti^0 in decreasing order of intensity. The shake up satellite of sputtered sample came down drastically due to reduction in concentration of TiO_2 on the surface. The high binding energy shoulder of O1s peak becomes sharper after sputtering due to removal of water adsorbed on the surface by ion beam. However intensity of this shoulder vanished at high binding energy side. The low binding energy main feature corresponds to Ti-O bond in TiO_2 while the high binding energy shoulder adsorbed oxygen in the surface. The main peak exhibited a shift of 0.4 eV to high binding energy with respect to the main peak of the as deposited sample indicating oxygen vacancies on the surface of the film created by ion beam sputtering. X-ray excited Auger Electron spectroscopy has been used to investigate the Auger transition $L_3M_{23}M_{23}$ and $L_3M_{23}M_{45}$ in Ti. Auger transitions $L_3M_{23}M_{45}$ and $L_2M_{23}M_{45}$ of as deposited sample are broader while the intensity of Auger peaks increased due to removal of surface contaminants like adsorbed carbon and oxygen. The valance band spectra of TiO_2 consists of O 2p and Ti 3d levels with O 2p at higher binding energy but no shift is observed in Ti 3s, 3p and O 2s peaks after sputtering. The AFM topography of TiO_2/Si(111) as deposited films indicate the formation of discontinuous layer with the formation of voids with an average size of 80nm. The peak position of Sr and Ti are in 2^+ and 4^+ charge states respectively observed by XPS. The X-ray satellite peak of as deposited sample decreases drastically after sputtering. The core levels of Ba 3d, Ti 2p and O 1s of the bulk and thin film of $BaTiO_3$ were investigated by XPS using AlK_α radiation. The results indicate that the core levels of

$BaTiO_3$ are broader compared to the core level of the thin film $BaTiO_3$. The observed binding energy shift of the core level in the thin film sample is related to two dimensional nature of the sample. The peaks in bulk sample are broader due to oxygen deficiency as the sample was prepared without oxygen flow whereas the thin film was prepared under oxygen partial pressure. It becomes defect free and gives rise to sharper peak. AFM topography of $BaTO_3$ shows formation of nano structure.

REFERENCES

[1] D.W. Richerson, Modern ceramic engineering: Properties, processing, and use in design, Second Edition, Marcel Dekker,inc., (1992).

[2] G.S. Oehrlein, F.M. d'Heurle and A. Reisman, J. Appl. Phys., 55(10)(198) 3715-3525.

[3] R.M. Wallace, G.D. Wilk, Crit. Rev. Solid State, 28 (2003) 231-285.

[4] G. Wilk, R.M. Wallace, J.M. Anthony, J. Appl. Phys., 89 (2001) 5243-5275.

[5] A. I. Kingon, J. P. Maria, Nature, 406 (2000) 1032-1038.

[6] J. Robertson, J. Vac. Sci. Technol. B, 18 (2000) 1785-1791.

[7] J.D. Plummer, P.B. Griffin, Proc. IEEE, 89 (2001) 240-258.

[8] M.A. Gribelyuk et al., J. Appl. Phys., 92 (2002) 1232-1237.

[9] M. Copel, M. C.Reuter, P. Jamison, Appl. Phys. Lett., 85 (2004) 458-460.

[10] M. Copel, Appl. Phys. Lett., 82 (2003) 1580-1582.

[11] K.B. Zhou and Y.D. Li, Angew. Chem. Int. Edn, 51 (2012) 602-13.

[12] O. Carp, L.C Huisman and A. Reller, Prog. Solid State Chem., 32 (2004) 33-177.

[13] M. Wang, P. Chen, R. Humphry-Baker, M.S. Zakeeruddin and M. Gratzel, Chem. Phys. Chem., 10 (2009) 290-299.

[14] A Navrotsky, Chem. Phys. Chem., 12 (2011) 2207-2215.

[15] J.-C. Sin, S.-M. Lam and A. Mohamed, Korean J. Chem. Eng., 28 (2011) 84-92.

[16] L.A. Brook, P. Evans, H.A. Foster, M.E. Pemble, A. Steele, D.W. Sheel,H.M, Yates, J. of photochem. and photobio A: Chemistry, 187 (2007) 53-63.

[17] A.A. Levchenko, G. Li, J. Boerio-Goates, B.F. Woodfield, A.Navrotsky, Chem. Mater., 18 (2006) 6324-6332.

[18] M. Osada, Y. Ebina, H. Funakubo, S. Yokoyama, T. Kiguchi, K. Takada, T. Sasaki, 18 (2006) 1023-1027.

[19] K. Seigban, C.N. Nordling, A. Fahlman, R. Nordberg, K. Hamrin, J. Hedman et al., Almqvist and Wiksells, Uppsala (1967).

[20] J.G. Jenkin, R.C.G. Leckey and J. Liesegang, J. Electron Spectrosc., 12 (1977) 1-35.

[21] M.K. Wu, J.R. Ashburn, C.J. Torng, P.H. Hor, R.L. Meng, L.Gao, Z.J. Huang, Y.Q. Wang and C. W. Chu, Phys. Rev. Lett., 58 (1987) 908-910.

[22] D.B. Geohegan in Laser Ablation of Electronic Materials :Basic Mechanisms and Appplications, E. Fogarassy and S. Lazare Eds. (North Holland, Amsterdam, 1992) P73.

[23] B. Dam, B. Stauble-Pumpin, J. Mat. Sci.: Mat. in Elec., 9 (1998) 217-226.

[24] Dave H.A. Blank, Guus J.H.M. Rijnders, Gertjan Koster and Horst Rogalla, Appl. Surf. Sci. 138-139 (1999) 17-23.

[25] M. Tyunina, J. Levoska and S. Leppavuori, J. Appl. Phys., 83 (1998) 5489-5496.

[26] T. Okada, Y. Nakata, M. Maeda, W.K.A. Kumuduni, J. Appl. Phys. 82 (1997) 3543-3547.

[27] J. Cheing, J. Horwitz, MRS Bullet in XVII (1992) 30-36.

[28] Raymund Kwok, xpspeaks4.1 software(Online).

[29] I.R. Lewis, H.G.M. Edwards, Handbook of Raman spectroscopy Marcel Dekker, New York, 2001.

[30] Characterization of Materials edited by E.N.Kaufmann , (John-Wiley and Sons publications,2003).

[31] D. Paramanik and S. Varma, J. Nanosci. and Nanotech., 07 (2007)2197-2200.

[32] D. Paramanik and S. Varma, J. Appl. Phys., 101(2007) 023528(1-7).

[33] G. X. Liu, F. K. Shan, W. J. Lee and B. C. ShinJ Korean Phys. Soc., 50 (2007) 1827.

[34] M. Walczak, M. Oujja, M. Sanz, J.F. Marco, M. Castillejo, Appl. Phys. A, 93 (2008) 735.

[35] J.F. Marco, A. Cesta, M. Gracia, J.R. Gancedo, P. Panjan, D. Hanzel, Thin Solid Films, 492 (2005) 158-165.

[36] D. O. Scanlon et al., Nature Materials, 12 (2013) 798.

[37] K.S. Kim and N. Winograd, Chem. Phys. Lett., 31 (1974) 312.

[38] V.R.R. Medicherla, R.R. Mohanta, Nimai C. Nayak, S. Choudhary, V. Soalnki, S. Varma, D. M. Phase, V. Sathe, AIP Conf. Proc., 1461 (2012) 403-407.

[39] R.R. Mohanta, V.R.R.Medicherala, K.L. Mohanta, Nimai C. Nayak, S. Ma-jumder,V. Solanki, Shikha Varma, Komal Bapna, D.M. Phase, V. Sathe ; Applied Surface Science, 325 (2015) 185-191.

[40] Mark C. Biesingera, Leo W.M. Laua, Andrea R. Gerson, Roger St.C. Smart, Appl. Surf. Sci, 257 (2010) 887-898.

[41] T. Choudhary, S. O. Saied, J. L. Sullivan and A.M. Abbot, J. Phys. D: Appl. phys., 22 (1989) 1185.

[42] Seif A. Nasser, Appl. Surf. Science, 157 (2000) 14-22.

[43] Sanjiv Kumar, V.S. Raju and T.R.N. Kutty, App. Surf. Science, 206 (2003) 250 -261.

[44] R.R. Mohanta, V.R.R.Medicherala, K.L. Mohanta, Nimai C. Nayak, S. Majumder, V. Solanki, Shikha Varma, Adv. Sci. Lett. 20 (2014) 584-587.

CHAPTER SIX

ANTIDIARRHOEAL ANTI-ULCER ACTIVITY OF DIETARY FIBER

Ashutosh Mishra[1*] & Manas Ranjan Mishra[2]

[1]*Department of Pharmacy, ITM GIDA, Gorakhpur, UP.*
**ashumusafer@gmail.com*
[2] *Department of Pharmacognosy, Gayatri College of Pharmacy, Sambalpur, Odisha.*

ABSTRACT

The dietary fiber (DF) is a complex mixture of carbohydrate polymers associated with a number of other, non-carbohydrate components. The DF is predominantly found in plant cell walls and consists of non-starch polysaccharides (NSP) together with lignin, protein, fatty acids, waxes, etc. Recently many studies on the effect of dietary fiber have highlighted their role in digestive physiology and their protective effect in gastrointestinal tract. The mechanism by which dietary fiber works on digestive system appears to be mediated by its soluble fiber that causes a significant enhancement in production and absorption of short chain fatty acids (SCFAs) in ileum. These absorbed short chain fatty acids can also have specific effects on intermediate metabolism. The insoluble fibers increase fecal bulk and, therefore, dilute its contents, which decreases interactions present in the faeces. Furthermore, insoluble fibers reduce intestinal transit time, allowing less opportunity for fecal mutagens to interact with the intestinal epithelium. In vivo antidiarrhoeal activity, onset of diarrhoea are more significantly increased in soluble dietary fiber in comparison to insoluble dietary fiber. Soluble dietary fiber has shown almost similar onset of diarrhoea as compared to the standard Loperamide. The overall findings has shown soluble dietary fiber as a powerful antidiarrhoeal agent. It may be due to its viscosity that binds the toxins and rehydrates the intestine by holding the water.The anti ulcer activity findings of soluble dietary fiber is due to its successful control of the ulcer induced by pyloric ligation. However, the insoluble dietary fiber has not shown any protective activity against induced ulcer.

Keywords: Dietary fiber, non-starch polysaccharides, viscosity, ulcer.

INTRODUCTION

The dietary fiber (DF) is a complex mixture of carbohydrate polymers associated with a number of other non-carbohydrate components. The DF is predominantly found in plant cell walls and consists of non-starch polysaccharides (NSP) together with lignin, protein, fatty acids, waxes, etc. [1].

The two types of dietary fiber are soluble dietary fiber (SDF) and insoluble dietary fiber (IDF). Most of the food contains mixture of both SDF and IDF. Pectin and gum are SDF found inside plant cells. They slow the passage of food through the intestines but do nothing to increase fecal bulk. Beans, oat bran, fruit and vegetables contain soluble fiber. In contrast, fibers in cell wall are IDF. These include cellulose, hemicellulose and lignin. Such fibers increase fecal bulk and speed up the passage of food through the digestive tract. Cereals, bran and whole grains along with vegetables contain insoluble fibers.

Table 1: Daily Intake of Dietary Fiber

Life Stage groups	Adequate intake (g/day)	
	Male	Female
1-3 years	19	19
4-8 years	25	25
9-13 years	31	26
14-18 years	38	26
19-30 years	38	25
31-50 years	38	25
51-70 years	30	21
Over 70 years	30	21
Pregnancy	-	28
Lactation	-	29

(Dietary Guidelines for Americans, 2005)

Cassia fistula L.

Cassia fistula L., (family: Caesalpiniaceae), a semi-wild Indian Labernum also known as the Golden Shower, are found as an ornamental tree for its beautiful bunches of yellow flowers in various countries including Mauritius, India, South Africa, Mexico, China, West Indies, East Africa and Brazil.

Cassia fistula seeds are rich in water-soluble gums (36.8%). All seeds of Cassia genera are found safe for internal use in various food formulations [2,3].

Tamarindus indica L.

Tamaridus indica L., (family: Caesalpiniaceae), a semi wild African plant, grows naturally in tropical and subtropical regions. The plant is identified as one of the most important plant resources for food materials [4].

Tamarindus indica seeds that can be peeled, roasted, or boiled are considered as famine food. Tamarind seeds contain polysaccharide (xyloglucans), which are extensively used as food thickeners, stabilizers and jelling agents in Japan. In the USA, its major industrial use has been as a wet end additive in the paper industry, replacement for starches and galactomannans [5]. According to Marathe et al. [6] the polyose from Tamarindus indica seed blends with pectin at different concentrations giving good jelly. It is also used more effectively in combination with other gums.

Definition of Dietary fiber

Hipsley [7] coined dietary fiber as a term for non-digestible constituents.

According to Trowell et al. [8], dietary fiber consists of the plant polysaccharides and lignin which are resistant to hydrolysis by digestive enzymes of man.

For Health and Welfare Canada [9] dietary fiber is the endogenous components of plant material in the diet which are resistant to digestion by enzymes produced by human. They are predominantly non-starch polysaccharides and lignin and may include, in addition, associated substances.

To U.S. Food and Drug Administration [10], dietary fiber is the material isolated by AOAC method 985.29.

A novel fiber source is a food that was manufactured to be a source of dietary fiber, and that (1) had not traditionally been used for human consumption to any significant extent, or (2) had been chemically processed (e.g., oxidized) or physically processed (e.g., finely ground) so as to modify the properties of the fiber, or (3) had been highly concentrated from its plant source [11].

Dietary fiber is substances of plant origin that cannot-be broken down to reabsorbable components by the body's own enzymes in the small intestine. Included are essentially soluble and insoluble non-starch polysaccharides (cellulose, pectin, hydrocolloids) and lignin and resistant starch. Substances like some sugar substitutes, organic acids, chitin and so on, which either are not or are incompletely absorbed in the small intestine, are not included [12].

Dietary fiber is the component of the food that is normally not broken down by the human body's own enzymes [13].

Dietary fiber is the edible substance of vegetable origin which is normally not hydrolyzed by the enzymes secreted by human digestive system [14].

American Association of Cereal Chemists (2000) -: Dietary fiber is the edible parts of plants or analogous carbohydrates that are resistant to digestion and absorption in the human small intestine with complete or partial fermentation in the large intestine. Dietary fiber includes polysaccharides, oligosaccharides, lignin and associated plant substances. Dietary fiber promotes beneficial physiological effects including laxation, and/or blood cholesterol attenuation, and/or blood glucose attenuation.

Dietary fiber is the material isolated by AOAC methods 985.29 and/or 991.43., combined with 997.08 [15].

Australia New Zealand Food Authority (Proposed) (2000)-: Dietary fiber is the fraction of the edible part of plants or their extracts, or analogous carbohydrates, that are resistant to digestion and absorption in the human small intestine, usually with complete or partial fermentation in the large intestine. The term includes polysaccharides, oligosaccharides (degrees of polymerization), and lignin's. Dietary fiber promotes one or more of these beneficial physiological effects: laxation, reduction in blood cholesterol, and/or modulation of blood glucose.

Table 2: Methods of Dietary Fiber Extraction

Reference (Method)	**Procedure Type**	**Measures**
Asp et al. (1983) [16]	Enzymatic-gravimetric	SDF, IDF, TDF
Craig et al. (2000) AOAC 2000.11 [17]	Enzymatic-ion exchange chromatographic	Polydextrose
Englyst and Cummings (1984) [18]	Enzymatic-gas chromatographic	TNSP
Englyst and Hudson (1987) [19]	Enzymatic-colorimetric	SNSP, by difference INSP from TNSP
Hoebregs (1997) AOAC 1997.08 [20]	Enzymatic-ion exchange chromatographic	Fructans

Lee et al. (1992) AOAC 991.43 [21]	Enzymatic-gravimetric using MES-TRIS buffer	SDF, IDF, TDF
McCleary et al. (2000) AOAC 999.03 [22]	Enzymatic-spectrophoto-metric	Fructans
Mongeau and Brassard (1993) AOAC 992.16 [23]	Enzymatic-gravimetric	SDF, IDF, TDF
Prosky et al. (1988) [24]	Enzymatic-gravimetric	TDF
Prosky et al. (1992) AOAC 991.42 [25]	Enzymatic-gravimetric	IDF
Prosky et al. (1994) AOAC 993.19 [26]	Enzymatic-gravimetric	SDF
Schweizer and Wursch (1979) [27]	Enzymatic-gravimetric	SDF, IDF, TDF
Southgate (1969) [28]	Enzymatic-gravimetric	SDF, IDF, TDF
Theander and Aman (1979) [29]	Enzymatic-gas chromatographic	Insoluble neutral polysaccharides, Soluble neutral polysaccharide, soluble uronic acids, Klason lignin, TDF
Theander and westerlund (1986) [30]	Enzymatic-gas chromatographic	Insoluble neutral polysaccharides, Soluble neutral polysaccharide, soluble uronic acids, Klason lignin, TDF
Uppsala method of Theander et al. (1995) AOAC 994.13 [31]	Enzymatic-gas chromatographic	Neutral polysaccharide, Uronic acid, Klason lignin, TDF
Official methods, of analysis AOAC 985.29 (1997)	Enzymatic-gravimetric	TDF, IDF, SDF

Table 3: Commonly used Dietary fibers

Sources	
Fruits and Vegetables	Apple, Kiwi and pears, Oranges, Pineapple, Peaches, Mango, Grape, Date, Sugar beet, Sugarcane, Olive, Potatoes, Carrot, Cauliflower, Cocoa hull and Chile
Legumes and Seeds	Pea and bean, Peanut hull, Sunflower hull, and Mesquite pod
Cereal grains	Rice, Corn, Barley, Distiller's grain, Psyllium, and Coffee Bean
Mushroom	Sclerotia

Fig.1 *Cassia fistula* L. plant

Plant Review

Cassia fistula L.

Position in the plant kingdom

Order -Rosales

Family - Caesalpiniaceae

Genus -Cassia

Fig.2 *Tamarindus indica* L.Plant

Species -fistula

Tamarindus indica L.

Position in the plant kingdom

Order - Fabales

Family - Caesalpiniaceae

Genus - Tamarindus

Species - indica

METHODOLOGY

Table 4: Collection of Seeds

Plant name	*Part used*	*Collection period*	*Source*	Mode of drying
Cassia fistula L.	Seed	December-January (2005)	BIT Mesra Campus, Ranchi, India	Oven drying at 60°C±5°C (48 h)
Tamarindus indica L.	Seed	April-May (2005)	BIT Mesra Campus, Ranchi, India	Oven drying at 60°C±5°C (48 h)

Flow Chart of Dietary Fiber Isolation

Sample Preparation and Digestion

Fat extraction (Pt. Ether 60-80°C)

↓

One gram of powder dissolved in 50 ml phosphate buffer (0.2M)

↓

Added heat stable 50 µl α-amylase (incubate at 95-100°C water bath for 30 min with agitation,) ↓

Added 10 ml 0.2 N NaOH and stirring (adjust pH 7.5-7.9 by adding 0.2 N HCl or 0.2 N NaOH) ↓

Added 100 µl protease solution and incubation for 30 min at 60±1°C with continuous agitation ↓

Added 10 ml 0.2 N HCl and stirring (adjust pH 4.0-4.7 by adding 0.2 N HCl or 0.2 N NaOH) ↓

Added 100 µl amylogucosidase solution while stirring (incubate 30 min at 60±1°C with constant agitation)

Digested samples were added in 90% methanol at 60°C

Removed from water bath covering with aluminum foil, it was left for 1 h for precipitation

↓

Filtered through Buckner funnel supported with whatsman filter paper, fitted the vacuum

↓

Washed residue 2 times each with 15 ml portion of 78%, 90% Methanol and acetone

↓

Used one sample for determination of protein and another for Ash value

↓

Enzyme digested sample were centrifuged

Washed 2 times with 10 ml distilled water at 70°C temperature

Washed two times with 78%, 90% methanol and acetone and reserved for determination of Soluble Dietary Fiber

↓

Used duplicate sample for analysis of protein and ash

↓

All Filtrate from Insoluble Dietary Fiber taken in beaker

↓

Added 4 volumes of 90% methanol preheated to 60°C

↓

Covered with aluminum foil, it was left for 1h for precipitation of soluble fiber

Used duplicate sample for analysis of protein and ash

C. fistula soluble dietary fiber *C. fistula*
insoluble dietary fiber

T. indica soluble dietary fiber *T. indica*
insoluble dietary fiber

Fig.3 Photographs of isolated dietary fiber (30x).

Dietary fiber effect on Gastro intestinal tract

Recently many studies on the effect of dietary fiber highlighted their role in digestive physiology and their protective effect in GI tract [32-34]. According to Aldoori et. al., 1997, soluble fiber-rich diets, reduce the incidence of duodenal ulcers in human. Dietary fibers may provide duodenal and gastric protection in populations that have higher fiber consumption [35]. Dietary fibers decrease gastric acid secretions [36], inhibit gastric emptying [37] and stimulate mucus cell proliferation. The mechanism by which dietary fiber works on digestive system appears to be mediated by its soluble fiber which causes significant enhancement in production and absorption of short chain fatty acids (SCFAs) in ileum. These absorbed short chains fatty acids can also have specific effects on intermediate metabolism. The insoluble fibers increase fecal bulk and, therefore, dilute its contents, which decreases interactions present in the faeces. Furthermore, insoluble fibers reduce intestinal transit time, allowing less opportunity for fecal mutagens to interact with the intestinal epithelium [38]. The soluble fibers have more prominent effect on stomach as compared to insoluble fiber. Previous studies have established that swelling rate of soluble fibers resulting in increased volume and hence results in distension of stomach affecting satiety and reduction in the rate of gastric emptying.

Considering these benefic ial effects ofdietary fiber on the gastrointestinal tract, the study is aimed to evaluate a possible protective action of fibers of C. fistula and T. indica against pyloric legated gastric ulcer and diarrhoea induced by castor oil in rats.

Antidiarrhoeal Activity

Antidiarrhoeal activity of dietary fiber was done on the castor oil induced diarrhoea.

Materials and Methods

Materials

Animals

Thirty inbreed Albino Wister rats (180-200g) of either sex maintained in animal house of BIT Mesra Ranchi were used. The animals were acclimatized for seven days under standard husbandry conditions maintained at a room temperature of (25 ± 2°C); relative humidity 45-55% and under 12:12 hours light/ dark cycle. The animals had free access to standard rat pellet and water was allowed ad libitum under strict hygienic conditions. Each experimental group had separate sets of animals and care was taken to ensure that animals used for one response were not employed elsewhere. Animals were allowed to get habituated to laboratory conditions for 48 hours prior to beginning of experimental protocol to minimize, if any, non-specific stress. All the protocols were approved by the Institutional Animal Ethical Committee (IAEC) of Department of Pharmaceutical Sciences; BIT Mesra, Ranchi, India.

Drugs, Chemicals and Equipments:

Dietary Fibers isolated by AOAC method (1996), Castor oil (Yogesh Pharma, Nanded, India), Loperamide (Dr. Reddy's labs, Hyderabad, India) were used in this study. Other chemicals used in the study were of analytical grade.

Methods

Treatment Protocol and method for castor-oil induced Diarrhoea in Rats:

Soluble dietary fiber was dissolved in distilled water and insoluble fiber in 2% w/v Tween 80 for oral administration at a concentration of 200, 400 mg/Kg body weights (bw).

All the animals were deprived of food and water, 24 h prior to the start of the experiment. Treatments of the various groups of animals are given below:

Group-I	Served as vehicle control and received 2% w/v Tween 80 at a dose 10ml/kg bw p.o.
Group-II	Received Loperamide 3mg/kg bw i.p. (Positive Control).
Group-III	Received Soluble Fiber-200 mg/Kg.
Group-IV	Insoluble Fiber-400mg/Kg (dissolve in Tween 80(2%w/v).

Castor-oil induced diarrhea 39-42].

Rats were divided into eight groups (n = 6) and, were fasted for 18 h and water was provided ad-libitum. The Dietary Fibers (200 and 400 mg/kg, p.o.) were administered orally to the first six groups of rats. One group received 10 ml/kg 0.5% v/v aqueous Tween 80 and served as a negative control. Another group received the standard drug Loperamide (3 mg/kg, p.o.) as positive control. After 1 h of treatment, all the animals were challenged with 1 ml of castor oil orally, by gavage and observed for consistency of faecal material. The frequency of defecation was noted in transparent plastic dishes placed beneath the individual rat cages upto 4h [43].

Statistical analysis

The results were expressed as mean±SD. Statistical significance between groups were assessed by One-way analysis of variance followed by Dunnett multiple test for the determination of level of significance.

ANTI-ULCER ACTIVITY

The antiulcer activity of dietary fiber was performed on the Wister rats pylorus ligation induced ulcer model.

Materials and methods

Animals:

As per the antidiarrhoeal activity.

Drugs, Chemicals and Equipments:

Dietary fibers isolated from *Cassia fistula* and *Tamarindus indica*, Ranitidine (Micro lab, Bangalore), Indomethacin (Aster Zeneca, Bangalore), 0.01N NaOH, Phenolphthalein, Topfer's reagent (dimethyl-amino-azo-benzene with Phenopthalin).

Pylorus ligation induced ulcer model:

Acute pylorus ligation in the rat causes copious secretion of acid. This acid responds to pylorus ligation and is probably elicited by vago-vagal reflexes activated by pressure receptors located in the pyloric gland area. This pylorus ligation stimulates the acid output over a time period of 4-6 hr. After that time there is no evidence of stimulated acid secretion. This hyper acidity is responsible for the induction of gastric ulcer on pylorus ligated rats [44, 45].

Treatment Protocol and method for pylorus ligation:

Soluble Dietary Fiber was dissolved in distilled water at a concentration of 100 and 200 mg/kg body weight, and IDF was dissolved in 2% Tween 80 at a concentration of 200 and 400mg/kg body weight.

All the animals were deprived of food and water 24 hours prior to the start of the experiment. Treatments of the various groups of animals are given below:-

Group-1	Served as vehicle control and received 2% Tween 80 w/v at a dose 2ml/kg bw p.o.
Group-2	Received ranitidine 20 mg/kg bw p.o. (Positive Control).
Group-3	Received *Cassia fistula* SDF at a concentration 200 mg/kg bw p.o.
Group-4	Received *Cassia fistula* ISF at a concentration 400 mg/kg bw p.o.
Group-5	Received *Tamarindus indica* SDF at a concentration 200 mg/kg bw
Group-6	Received *Tamarindus indica* IDF at a concentration 400 mg/kg bw p.o.

Procedure for Pylorus Ligation:

Rats were fasted for 24 h and given water *ad libitium* in individual cages. Care was taken to avoid coprophagy. DF, ranitidine, and tween 80 were administered 30 minutes prior to pyloric ligation to the respective groups. Under light ether anesthesia, the abdomen was opened and the pylorus was ligated. The abdomen was then sutured. After 4 h of ligation, the animals were sacrificed and the stomach was dissected out. The gastric content was collected and the volume was measured. The stomach was then opened along the greater curvature and the ulcerative index (UI) was determined by examining the inner lining of each stomach [44, 46].

Assessments of ulcer index:

Gastric ulceration was assessed by means of "Scoring Technique" because the ulcer lesion varied with increasing frequency from deep linear ulcers that are occasionally in excess of 8 mm in length to small circular ulcer up to 2 mm in diameter. The lesion induced by pylorus ligation and indomethacin were found in the gastric areas and glandular region of the stomach. The severity and extent of gastric damage was scored on visual basis and for the more accurate scoring, the microscopic examination of the dissected stomach was carried out with hand lens with 2x magnification. The method of scoring was a modification of the method of in which we decreased the number categories scored [47].

Scoring of ulcer:

Lesion	Scores	
	Each	Maximum
Deep linear ulcers > 8 mm length	4	4
Deep linear ulcers 4–8 mm length	3	6
Deep linear ulcers ≤ 4 mm length	2	8
Circular ulcers 1–2 mm in	1	–
	0.5	–

diameter Circular ulcers 1–2 mm in diameter Fraction of stomach showing evidence of hemorrhage	2	2

$$\text{The mean ulcer index was determined by} = \frac{\text{Total ulcer in each group}}{\text{Total number of animal in each gruop}}$$

Analysis of gastric contents:

The gastric contents of each stomach were analyzed individually and volume of gastric contents measured (whenever volume of recovered gastric secretions was sufficient). The specimens were centrifuged in a graduated centrifuge tube at 2000 revolution per minute for 10 minutes. The volumes of supernatant and of solids were recorded. The supernatant was removed and Hydrogen ion concentration was determined using pH paper. The free acidity was assayed by titration to pH 3.5 with 0.01 N sodium hydroxide using Topfer's reagent as indicator and total acid production by titration to pH 8.0 with 0.01 N sodium hydroxide using phenolphthalein as an indicator as follows:

1 ml of filtered gastric contents was pipetted into a small beaker (60 ml volume). 2-8 drops of Topfer's reagent was added which gave red colour to pipetted gastric solution and then it was titrated with 0.01 N sodium hydroxide until all traces of the red colour disappeared and the colour changed to yellowish orange. The volume of alkali added was recorded. This volume was free

acid volume. Then 2-3 drops of phenolphthalein was added and titration was continued until a definite red tinge reappeared. Again the burette was read for the total volume of alkali added. This volume measured combined acidity. The sum of free and combined acidity volumes gave total acidity volume. Appearance of yellow colour on adding Topfer's reagent signifies absence of free acid. Observations were expressed in mEq/L by multiplying factor 10 with observations recorded in ml.

Acidity = Volume of NaOH×Normality×100 mEq/L/100 g of body weight.

Histopathological Studies

The dissected stomachs from both the models were preserved in 10% formalin and processed for routine paraffin block preparation. Using an Optical Rotary microtone, section of thickness about 5μm were cut and stained with hematoxylin and eosin (H&E). These were examined under microscope for histopathological changes such as congestion, corrosion haemorrhage and necrosis [48, 49].

Statistical analysis:

The results were expressed as mean±SD. Statistical significance between groups were assessed by One-way analysis of variance followed by Dunnett multiple test for the determination of level of significance.

DISCUSSION

Soluble fibers are more significant in controlling diarrhoea induced by castor oil than insoluble fiber. CSF showed no onset diarrhoea and low number wet stool even after 4h but it shows the total faeces number 13.33±5.03 and total pellets weights 3.61±0.19 g. TSF showed diarrhoea at 196±11.50 min, total number of feces 10.00±2.00 g, wet faeces 3.67±0.58 g, total weight of feaces 2.61±0.18 g. CIF and TIF had showed no promising result against diarrhoea in comparison of Guar gum and Tragacanth SDF more prominently reduce the diarrhoea and gastric lesions because of its viscosity by which it hydrated the intestine and bind the excess acid and neutralize it. Most SDF are used as an energy source of bacteria and converted to various substances. When the intestinal flora was investigated, the soluble dietary fiber administered was found to decrease significantly the number of aerobic bacteria, thus making anaerobic bacteria predominant in the intestinal flora. In general, anaerobic bacteria are thought to play a protective role in intestinal flora. Therefore, the administration of the fiber may help normalize the intestinal flora. Moreover, SCFAs such as acetic acid, propionic acid, and butyric acid are produced mainly by enteric bacteria in the process of digesting dietary fiber. These fatty acids comprise approximately 80% to 90% of the overall SCFAs produced in the digestive tract.

More than 0.7 of total CO_2 produced by colonocytes is derived from the SCFA n-butyrate and has been convincingly shown to stimulate sodium absorption. These effects of butyric acid is advantageous, both in healthy persons and especially in the hydration therapy in various cases of diarrhoea, where butyric acid could provide the energy to the colonic mucosa thus keeping it active, and promote the water and sodium absorption.

Table 5 Effect of Dietary fibers in castor oil induce diarrhoea on rats

Treatment	Dose	Onset time of diarrhea (min)	Total number of faeces	Number of wet faeces	Total weight of faeces (gm)
Control	Castor oil 10 mg/kg	55.67±5.68	19.33±2.52	14.67±1.53	4.87±0.356
Standard	Loperamide 4mg	245.67±12.50	3.33±0.58	1.33±0.58	1.19±0.02
Guar Gum	200 mg/kg	206.00±9.17	11.33±3.05	4.67±1.16	3.09±0.26
Tragacanth	200 mg/kg	83.33±5.69	14.00±2.00	9.33±1.53	3.66±0.43
CSF	200 mg/kg	276.35±12.50	6.33±3..03	1.33±0.58	2.31±0.44
CIF	400 mg/kg	76.33±12.50	12.00±4.00	8.00±1.00	3.61±0.19
TSF	200 mg/kg	196.33±11.50	10.00±2.00	3.67±0.58	2.61±0.18
TIF	400 mg/kg	76.67±4.10	13.67±4.50	8.33±2.08	3.63±0.29

The values represent the means ±for six rats per group.

CSF, *Cassia fistula* soluble fiber. CIF, *Cassia fistula* insoluble fiber. TSF, *Tamarindus indica* soluble fiber. TIF, *Tamarindus indica* insoluble fiber.

P <0.05, ** *P* <0. 01 and *** *P* <0. 001 compared to control

Results are expressed as mean ± S.E.M., n = 6. Statistical significance test with control was done by student 'T' test. *P < 0.05, **P< 0.01, *** P<0.001 when compared to control.

The effect of CSF, TSF, CIF and TIF treatment on pyloric legated induced gastric ulcers in rats are shown in Table 6. Ulcer protection effect TSF showed more significant result in comparison to CSF. TSF have more ulcer protective effect, 82.32% in respect of CSF 55.93% and standard ranitidine 68.35% as compared to control. Volume of gastric juice and free acid of both the fibers were significantly lower than the control; however, IDF showed no significant result on the protection of ulcer against the pyloric legated rats.

In histopathology, control group showed marked edema of the *lamina propria* with infiltration of neutrophils above the basement membrane as well as in-between the mucosal glands. There was desquamation of superficial epithialal cell in large areas. However, at some places the damage and erosion of mucosa was very severe leaving only intact basement membrane and lamina propria. Congestive and haemorrhagic changes were also seen at few places in lamina propria.

Soluble fibers showed mildest form of changes. The lamina propria exhibited very mild edema and slight vascular congestion as compared to control group. The muscular layer remained intact with sulemu cesal layer. Very mild focal superficial mucosal disruptions with along scattered neutrophils were present among the surface epithelial layer and lumen of mucosal glands. Congestive and haemorrhagic changes were also very mild as compared to control group.

Microscopic observations in rat treated with insoluble fiber showed more severe or similar degree of haemorrhagic changes than the degree of changes found in case of control rats. The erosion led to complete loss of mucosal in a wide area leaving only muscularis mucosis. The edema of lamina propria and submucosal was also severe, which widely separated the mucosal and muscularis layer. There was a marked acute inflammatory infiltration and extrusion of fibrin in edematous places.

The activities of soluble and insoluble fibers were not clearly distinguished in free acid content and its pH, but it has very huge gape in percentage of ulcer protective effect. The *Tamarindus indica* soluble fiber produced promising result in comparison of control. Dietary fiber is best considered as a dietary concept that integrates its physiological and physical functions, and its diverse effects, both locally and systematically to the gastrointestinal tract. Research on new gastroprotective agents include the use of different models of experimentally induced gastric lesions [50]. Edematous and discomative changes can be attributed to less blood supply, increase hydrostatic pressure and acid secretion with backflow, regurgitation of detergent bile salts and lysolesithin due to parastolic movement. The ulcer protective effect of the soluble part may be attributed due to binding and adsorbing to H^+ ion, bile acid and lysolesithin. However, the insoluble have some irritant absorption.

Table 6 Effect of Dietary fibers on pyloric legated rats induced ulcer gastric contents and ulcer score

Group	Design of treatment	Dose mg/kg bw	Volume of gastric juice ml	pH	Free acid (meq/L)	Ulcer score	% of ulcer protection
1	Control (normal saline)	2	10.32±0.39	2.04±0.31	24.90±0.41	3.79±0.27	0.00
2	Ranitidine	20	4.15±0.28**	4.74±0.28**	20.58±0.53**	1± 0.40**	65.41±2.73**
3	CSF	200	6.43±0.21**	4.17±0.29**	19.20±0.20**	1.67±0.29**	56.82±0.67**
4	CIF	400	5.07±0.42**	3.83±0.29**	15.37±0.25**	4.33±0.29	-12.63±1.74
5	TSF	200	6.13±0.23**	3.67±0.29**	13.80±0.40**	0.67±0.29**	78.92±3.857**
6	TIF	400	5.13±0.30**	4.50±0.50**	12.73±0.40**	5.0±0.50	29.99±2.26

The values represent the means ± SD for six rats per group.

CSF, *Cassia fistula* soluble fiber. CIF, *Cassia fistula* insoluble fiber. TSF, *Tamarindus indica* soluble fiber. TIF, *Tamarindus indica* insoluble fiber.

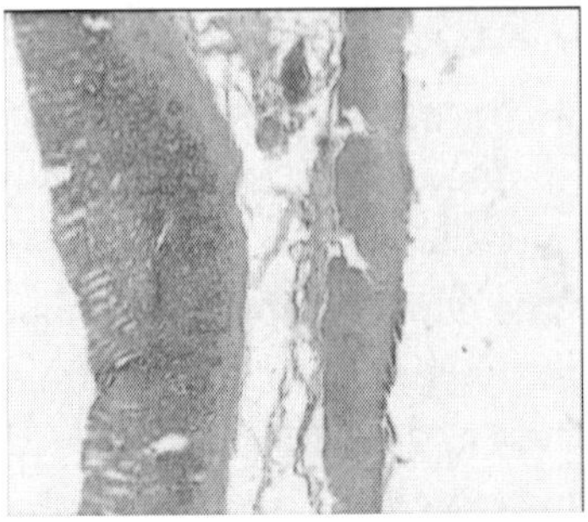

FIG a: Vehicle Control

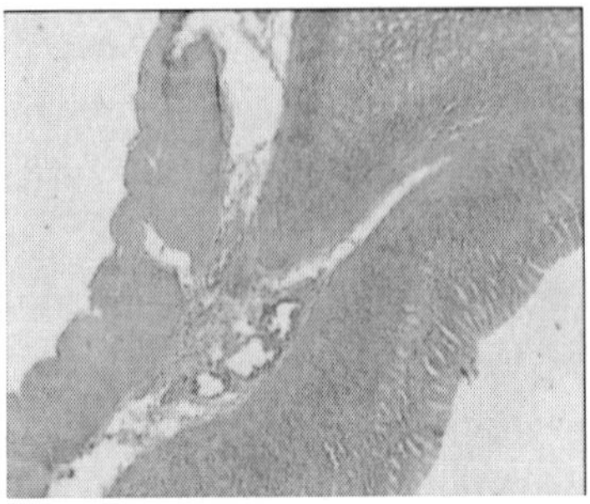

FIG b: CIF

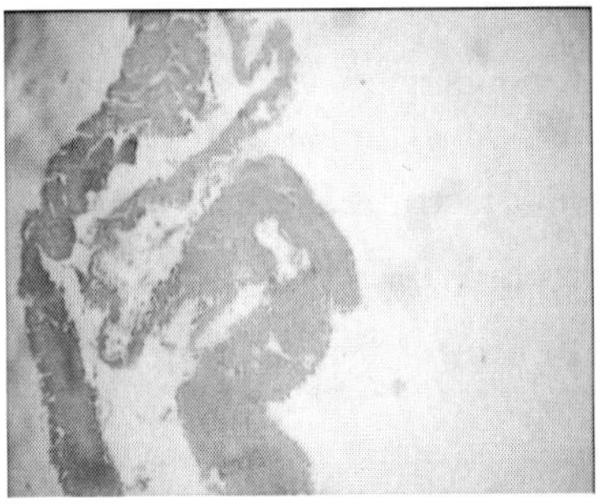

FIG c: TIF

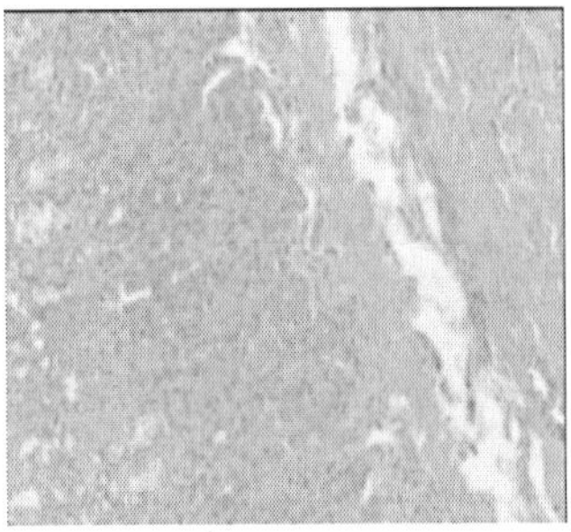

FIG d: TSF

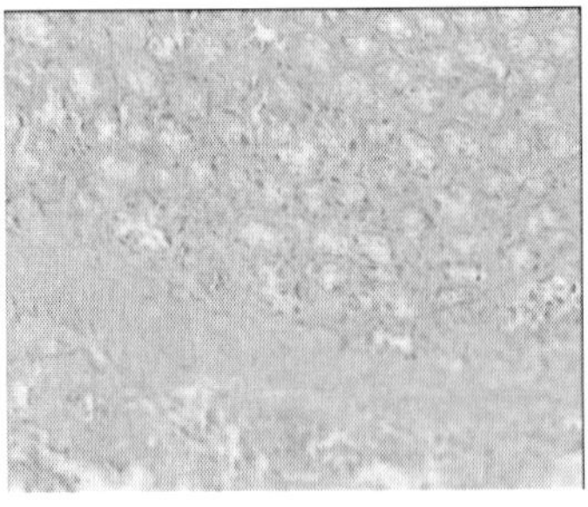

FIG e: CSF

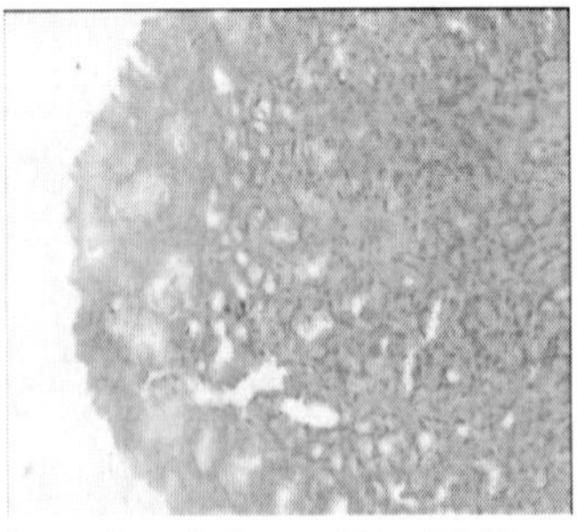

FIG f: Standard drug Ranitidine

Fig.4 Section through gastric mucosa of stomach. Hematoxylin and eosin, 10x and 40X; CSF, *Cassia fistula* soluble fiber. CIF, *Cassia fistula* insoluble fiber. TSF, *Tamarindus indica* soluble fiber. TIF, *Tamarindus indica* insoluble fiber.

CONCLUSION

In vivo antidiarrhoeal activity, onset of diarrhoea are more significantly increased in soluble dietary fiber in comparison to insoluble dietary fiber. Soluble dietary fiber has shown the almost similar onset of diarrhoea as compared to the standard Loperamide. The overall finding has shown that soluble dietary fiber is powerful antidiarrhoeal. It may be due to its viscosity, by which it binds the toxins and rehydrates the intestine by holding the water.

The anti ulcer activity finding has shown that the soluble dietary fiber successfully controls ulcer induced by pyloric ligation. However, the insoluble dietary fiber has not shown any protective activity against induced ulcer.

REFERENCES

[1] McDougall G J, Morrison I M, Stewart D, Hillman JR (1996) Plant cell walls as dietary fiber: Range, structure , processing and function. J Sci Food Agric 70: 133-150.

[2] Hallagan JB, Ladu BN, Pariza MW, Putnam JM, Borzelleca JF (1997) Assessment of Cassia Gum. Food and chemical toxicology 35: 625-632.

[3] Kapoor VP, Taravel FR, Joseleau J-P, Milas M, Chanzy H, Rinaudo M (1998) *Cassia spectabilis* DC seed galactomannan: Structural, crystallographical and rheological studies. Carbohydr Research 306: 31-241.

[4] Yusuf AA, Mofio BM, Ahmed AB (2007) Proximate and mineral compositin of tamarindus indica linn 1753 seeds. Science World J 2: 1-4.

[5] Perprtuo GF, Salgado JM (2003) Effect of mango (*Mangifera indica*, L.) ingestion on blood glucose levels of normal and diabetic rats. Plant Foods for Human Nutr 58: 1-12.

[6] Marathe RM, Annapure US, Singhal RS, Kulkarni PR (2002) Gelling behavoiur of polyose from tamarind kernel polysaccharide. Food Hydrocolloids 16: 423-426.

[7] Hipsley EH (1953) Dietary "fibre" and pregnancy toxaemia Br Med J 2: 420-422.

[8] Trowell HC, Southgate DAT, Wolever TMS, Leeds AR, Gassull MA, Jenkins DJA (1976) Dietary fibre redefined. The Lancet 1: 967.

[9] Health and Welfare Canada (1985) Report of the Expert Advisory Committee on Dietary Fibre. Ottawa Supply and Services Canada.

[10] USFDA (U.S. Food and Drug Administration) (1987) Nutrition labeling of food calorie content Federal Register 52: 28590–28691

[11] Health Canada (1988) Guideline Concerning the Safety and Physiological Effects of Novel Fibre Sources and Food Products Containing Them. Ottawa: Food Directorate, Health Protection Branch, Health Canada.

[12] Anonymous (1989) GDch Stellungnahme der Untergruppe "Ballaststoffe" der Arbeitsgruppe "Freagen der Emahrung" der Fachgruppe "Lebensmittelchemie und gerichtliche Chemie" in der GDCh" Lebensmittelchemie und Gerichtliche Chemie 43: 113-117.

[13] Anonymous (1992) Belgian Food Law, KB 8/1/1992

[14] Anonymous (1993) Italian Food Law, DL 16/2/1993

[15] Hignett R (2000) Letter to All Interested Parties. Nutrition Labelling of Dietary Fibre. [Online]. Available: http://foodstandards.gov.uk/farm_fork/nutfibre. htm [accessed March 14, 2001].

[16] Asp N-G, Johansson C-G, Hallmer H, Siljeström M (1983) Rapid enzymatic assay of insoluble and soluble dietary fiber. J Agric Food Chem 31: 476-482.

[17] Craig SAS, Holden JF, Khaled MY (2000) Determination of polydextrose as dietary fiber in foods. J AOAC Int 83: 10061012.

[18] Englyst HN, Cummings JH (1984) Simplified method for the measurement of total non-starch polysaccharides by gas-liquid chromatography of constituent sugars as alditol acetates. Analyst 109:937-942.

[19] Englyst HN, Hudson GJ (1987) Colorimetric method for routine measurement of dietary fibre as non-starch polysaccharides. A comparison with gas-liquid chromatography. Food Chem 24: 63–76.

[20] Hoebregs H (1997) Fructans in foods and food products, ion-exchange chromatographic method: Collaborative study. J AOAC Int 80: 1029-1037.

[21] Lee SC, Prosky L, De-vries JW (1992) Determination of total soluble and insoluble die-tary fiber in foods—Enymatic-gravimetric method, MES-TRIS buffer: Collaborative study. J AOAC Int 75: 395-416.

[22] McCleary BV, Murphy A, Mugford D C (2000) Measurement of total fructan in foods by enzymatic/spectrophotometric method: Collaborative study. J AOAC Int 83: 356-364.

[23] Mongeau R, Brassard R (1993) Enzymatic-gravimetric determination in foods of dietary fiber as sum of insoluble and soluble fiber fractions: Summary of collaborative study. J AOAC Int 76: 923-925.

[24] Prosky L, Asp N G, Schweizer TF, DeVries JW, Furda I (1988) Determination of insoluble, soluble, and total dietary fiber in foods and food products: Interlaboratory study. J Assoc Off Anal Chem 71: 1017-1023.

[25] Prosky L, Asp N-G, Schweizer TF, DeVries JW, Furda I (1992) Determination of insoluble and soluble dietary fiber in foods and food products: Collaborative study. J AOAC Int 75: 360-367.

[26] Prosky L, Asp N-G, Schweizer TF, DeVries JW, Furda I, Lee SC (1994) Determination of soluble dietary fiber in foods and food products: Collaborative study. J AOAC Int 77: 690-694.

[27] Schweizer TF, Würsch P (1979) Analysis of dietary fibre. J Sci Food Agric 30: 613-619.

[28] Southgate DAT (1969) Determination of carbohydrates in foods. II. Unavailable carbohydrates. J Sci Food Agric 20: 331-335.

[29] Theander O, Åman P (1979) Studies on dietary fibres 1. Analysis and chemical characterization of water-soluble and water-insoluble dietary fibres. Swedish J Agric Res 9: 97-106.

[30] Theander O, Westerlund E (1986) Determination of individual components of dietary fiber. In: Spiller GA, ed. CRC Handbook of Dietary Fiber in Human Nutrition, Boca Raton, FL: CRC Press.

[31] Theander O, Åman P, Westerlund E, Andersson R, Pettersson D (1995) Total dietary fiber determined as neutral sugar residues, uronic acid residues, and Klason lignin (the Uppsala method): Collaborative study. J AOAC Int 78: 1030-1044.

[32] Naitoa Y, Takagib T, Katadab K, Uchiyamab K, Kurodab M, Kokuraa S, Ichikawab H, Watabec J, Yoshidaa N, Okanouea T, Yoshikawa T (2006) Partially hydrolyzed guar gum down-regulates colonic inflammatory response in dextran sulfate sodium-induced colitis in mice. J of Nutritional Biochem 17: 402-409.

[33] Vandamme TF, Lenourry A, Charrueau C, Chaumeil J-C (2002) The use of polysaccharides to target drugs to the colon. Carbohydr polymers 48: 219-231.

[34] Ebihara K, Masuhara R, Kiriyama S, and Manabe M (1981) Correlation between viscosity and plasma glucose- and insulinflattening activities of pectins from vegetables and fruits in rats. Nutr Rep Int 23: 985-992.

[35] Rydning A, Berstad A (1985) Fiber diet and antacids in short-term treatment of duodenal ulcer. Scand J Gastroenterol 20: 1078- 82.

[36] Aldoori WH, Giovannucci EL, Stampfer MJ, Rimm EB, Wing AL (1997) Willett WC., A prospective study of alcohol, smoking, caffeine, and risk of duodenal ulcer in men. Epidemiology 8: 420-4.

[37] Sousa, IMN, Matias EC, Laureano O (1997) The texture of low calorie grape juice jelly. Z Lebensm Unters Forsch A 205: 140-142.

[38] Thebaudin JY, Lefebvre AC, Harrinton M, Bourgeois CM (1997) Dietary fibres: Nutritional and technological intrest. Trends in food Sci and technol 8: 41-48.

[39] Capasso A (1994) Dissociation of castor-oil induced diarrhea and intestinal mucosal injury in rat, Br J of pharmacol 113: 1127-1130.

[40] Mascolo N, Izzo AA, Barbato F, Capasso F (1993) Inhabits of nitric oxide synthatase prevent castor-oil induced diarrhea in rats. British Journal of pharmacol 108: 861-864.

[41] Patel NJ, Gujarati VB, Gouda TS, Venkat N, Rai A, Nandakumar M, Shantakumar S (2006) Antidiarrhoeal activity of alcoholic and aqueous extracts of roots of *tylophora indica* in rodents, Pharmacol online 1: 19-29.

[42] Rani S, Ahmed N, Rajaram S (1999) Antidiarrhoeal evaluation of *Clerodendrum phlomidis* Linn. Leaf extract in rats. J Ethnopharmacol 68: 315-319.

[43] Kumar S, Dewan S, Sangrula H, Kumar VL (2001) Antidiarrhoeal activity of the latex of *Calotropis procera*. J Ethnopharmacol 76: 116-118.

[44] Vogel GH (2002) Drug Discovery and Evaluation–Pharmacological Assays CD-ROM, 2nd ed. Springer Verlag, Berlin.

[45] Kulkarni SK (1999) editor; pharmacology of gastrointestinal tract.Handbook of Experimental Pharmacology, 3rd ed. Delhi:Vallab Prakashan, pp 148-49.

[46] Ghosh MN (2005) Some common evaluation technique. In: Hilton & Company, editor. Fundamental of Experimental Pharmacology. 3rd ed Kolkata: S K Ghosh & Others, pp 180.

[47] Best R, Lewis DA, Naser N (1984) Studies of mechanism of Pylorus ligation induced gastric damage in rats. Gastroenterology 82: 107-116.

[48] Endo H, Yoshida H, Ohmi N, Higuchi S (2001) Effects of Lansoprazole and Amoxicillin on Uptake of [14C] Clarithromycin into Gastric Tissue in Rats. Antimicrob Agents Chemother 45: 3451-55.

[49] Alumets J, Ekelund M, Håkanson R, Hedenbro J, Rehfeld JF, Sundler F, Vallgren S (1982) Gastric acid response to pylorus ligation in rats: Is gastrin or histamine involved? J Physiol.323: 145-156.

[50] Winterbourn, C, Hawkins, RE, Brian, M, Carrell, RW (1975) The estimation of red cell superoxide dismutase activity. J Lab Clin Med 85: 337- 41.

CHAPTER SEVEN

STUDIES ON TRANSITION METAL COMPLEXES WITH MACROCYCLIC LIGANDS

Sujata Karna

School of Chemistry, Sambalpur University, Odisha.
sujatakarna@gmail.com

ABSTRACT

Macrocycles are important and powerful ligands, ubiquitous in transition metal coordination chemistry for the reasons that,they mimic important biological ligands developed long ago by nature, for example the porphyrin prosthetic group of many metalloproteins. They impart thermodynamic and kinetic stabilities to their metal complexes uncommon or non-existent with ligands of lessor types. For the purposes of this dissertation, a macrocycle is defined as a cyclic compound with nine or more members including at least three donor atoms. In this introduction, topologically constrained azamacrocycles, defined as macrocycles having additional bridges besides the basic parent ring between nitrogen atoms donors, will be discussed. Special attention will be paid to a unique class of macrocycles consisting of cross-bridged macrocyclic ligands having added components of topological and rigidity constraints.The present work relates to the investigation on the preparation of dinuclear acyclic complexes by the action of 3,6-dimethyl-2,7-dioxo-4,5-diazaocta-3,5-diene-2,7-dihydrazone (BAAD) on the metal ions. The open chain diacetyldihydrazone

(BAAD) metal complexeshave been prepared as precursors for the preparation of macrocyclic complexes. They have been characterized on the basis of their spectral, magnetic and thermo chemical properties which have been used as valuable tools in the elucidation of the structure.Here we describe some bis-chelate complexes of BAAD with the metal ions Co(II), Ni(II) and Cu(II).

Keywords: Macrocycles, metalloproteins, azamacrocycles, BAAD,

INTRODUCTION

Macrocycles are important and powerful ligands, ubiquitous in transition metal coordination chemistry for the following reasons. 1) They mimic important biological ligands developed long ago by nature, for example the porphyrin prosthetic group of many metalloproteins [1]. 2) They impart thermodynamic and kinetic stabilities to their metal complexes uncommon or non-existent with ligands of lesser types.

For the purposes of this dissertation, a macrocycle is defined as a cyclic compound with nine or more members including at least three donor atoms [2]. In this introduction, topologically constrained azamacrocycles, defined as macrocycles having additional bridges besides the basic parent ring between nitrogen atom donors, will be discussed. Special attention will be paid to a unique class of macrocycles consisting of cross-bridged macrocyclic ligands having added components of topological and rigidity constraints.

The chemistry of macrocyclic complexes has attracted the interest of both inorganic and bioinorganic chemists in recent years [3].The field of coordination chemistry of macrocyclic complexes has undergone remarkable growth during the past few decades and become a growing class of research [4]. This enormous growth is due to synthesis of a large number and variety of synthetic macrocycles which behave as coordinating agents for metal ions. Template reactions lie at the heart of macrocyclic chemistry and are the best aids for the preparation of macrocyclic complexes. Generally transition metal ions have been used as templates. The importance of macrocyclic complexes in coordination chemistry is largely because of its various applications in biological processes such as photosynthesis and dioxygen transport, catalytic properties, potential applications as metal extractants and radio therapeutic agents. The importance of macrocyclic complexes is also due to their resemblance with many natural systems like porphyrins and cobalamines [5,6]. Macrocyclic complexes have attracted attention because of their antibacterial and antifungal activities. Some macrocyclic complexes have been reported to have anti-inflammatory properties. Several macrocyclic complexes with tetraazamacrocyclic ligands, such as cyclam or bicyclam have been reported to exhibit antitumour activity. Macrocyclic metal complexes of lanthanides, eg, Gd^{+3} are used as MRI-contrast agents [7]. The chemistry of macrocyclic complexes is also important due to their use as dyes and pigments as well as NMR shift reagents.

Macrocyclic complexes have undergone a phenomenal growth during the past three decades. The impetus for research has been accentuated by the recognition of the fact that the basic structural units of several naturally occurring systems such as haemoglobin, chlorophyll, vitamin B_{12} and many others contain macrocycliccorrin, porphin or chlorine rings [1-3], which have assumed a key role in the biological evolution on our planet. These splendid ligands are commonly bound to metal ions like iron (II), magnesium (II) or cobalt (II). Studies encompassing syntheses, structures and reactivities on man-made and natural macrocycles have been fruitful for a profound understanding of the role of metal ions in biological systems. Metal phthalocyanines [8], discovered in 1934 and the macrocyclic compounds have attracted attention for their industrial applications, catalytic activity for syntheses of unusual organic compounds and their semiconductor behavior.

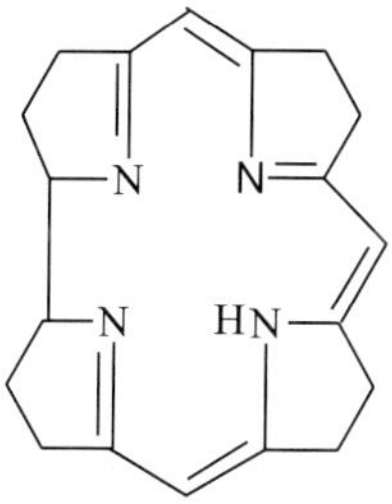

[1]

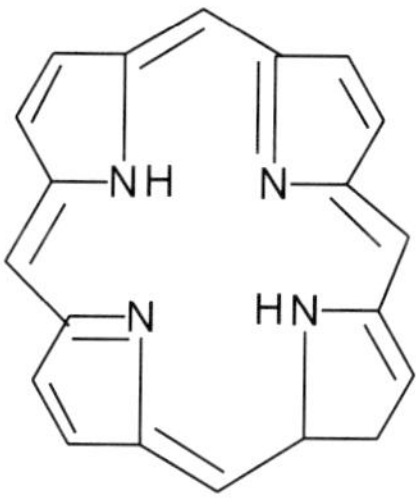

[2]

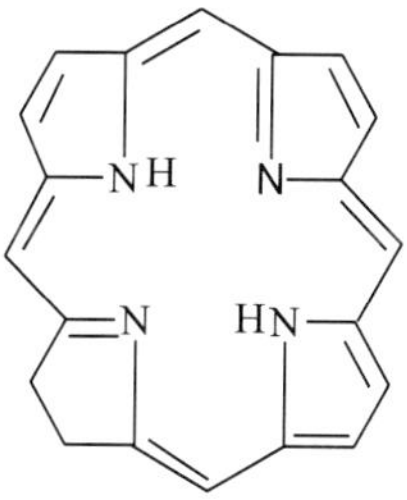

[3]

Macrocyclic metal complexes have received considerable attention in recent years as model intermediates in the elucidation of mechanisms of biological processes. Most of the studies deal with mononuclear complexes. Recently some binuclear macrocyclic complexes have also been synthesized using Schiff bases of polyketones [1], crown ethers [2] and some nitrogen donor ligands [3-5]. In the previous section a series of binuclear bis- complexes of Cobalt(II) and Copper(II) of the type

$M_2(BAAD)_2X_4$ [M=Co(II), X= Cl^-, Br^-, NO_3^-, M=Cu(II), X= Cl^-, Br^-, ClO_4^-] and $M_2(BAAD)_2X_4$ [M=Ni(II), X= Cl^-, Br^-, NO_3^-, ClO_4^-] with the ligand (BAAD), 3,6-dimethyl-2,7-dioxo-4,5-diazaocta-3,5-diene-2,7-dihydrazone have been described. The pendant $-NH_2$ groups of these hydrazone complexes are now suitably oriented to react with structurally different functions to yield macrocyclic complexes enclosing two metal centres. Report on the results of investigations on a series of binuclear macrocyclic complexes of Co(II), Ni(II) and Cu(II) with the cyclic ligand 3,4,7,8,14,15,18,19-octamethyl-1,2,5,6,9,10,12,13,16,17,20,21-dodecaazacyclo-11,22-dithia-2,4,6,8,13,15,17,19-octaene **(ODDO)** ring(4) formed by the in situ reaction of 2 mol of carbondisulphide with bis-BAAD complexes $M_2(BAAD)_2X_4$ is described in this section.

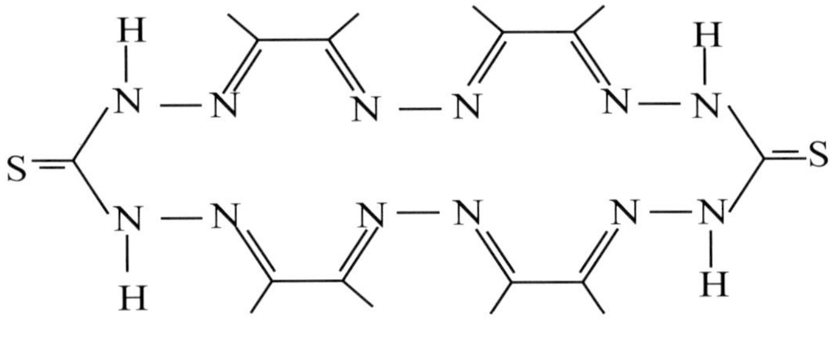

[4]

METHODOLOGY

Transition metal complexes with vicinal dihydrazones have assumed importance [9-15]. The bis and tris-(2,3-butanedionedihydrazone) metal complexes have been used to prepare mononuclear macrocyclic [16] and the three dimensional clathrochelate [9] metal complexes by the reaction of aldehyde and ketones. Amongst the dihydrazone ligands, 3, 6-dimethyl-2,7-dioxo-4,5-diazaocta-3,5-diene-2,7-dihydrazone(BAAD) possesses unique structural characteristics in relation to the position of nitrogen atoms and conjugated π-electron systems. As has been already stated, it can act as a tetra functional ligand coordinating with metal ions in a bi-bidentate manner. As a matter of fact this ligand has been used to give binuclear complexes, $[Fe_2(BAAD)_3]I_4$ and $[Ni_2(BAAD)_3]I_4$ [12]. A bi-bidentate coordination of the ligand with metal ions has been proposed in analogy with similar ligands such as pyridylaldazine [12].

The present work relates to the investigation on the preparation of dinuclear acyclic complexes by the action of BAAD on the metal ions. The open chain acetyldihydrazone metal complexeshave been prepared as precursors for the preparation of macrocyclic complexes. They have been characterized on the basis of their spectral, magnetic and thermo chemical properties which have been used as valuable tools in the elucidation of the structure. In the present chapter, we describe some bis-chelate complexes of BAAD with the metal ions Co(II), Ni(II) and Cu(II).

RESULTS AND DISCUSSION

Metal complexes of the ligand BAAD have been prepared in 1:2 stoichiometry with the composition $[M_2(BAAD)_2X_4]$ (M=Co(II),Ni(II) and Cu(II) and X=Cl^-, Br^-, NO_3^- and ClO_4^-. The analytical data of the metal complexes are presented in Table-1.The complexes are insoluble in most of the organic solvents but are moderately soluble in water, DMF and DMSO.

They are stable under normal conditions and are slowly attacked by alkalies and acids. The perchlorate complexes are sensitive to moisture and get decomposed when exposed to air. They are explosive in nature.

Table-1 Analytical data and Colour of BAAD complexes.

Complexes	Colour	Found (Calculated)%.		
		Metal	Anion	Nitrogen *
$[Ni_2(BAAD)_2Cl_4]$	Green	17.95 (18.09)	21.65 (21.77)	25.29 (25.76)
$[Ni_2(BAAD)_2Br_4]$	Pale green	14.16 (14.21)	38.45 (38.55)	19.91 (20.24)
$[Ni_2(BAAD)_2(NO_3)_4]$	Brownish green	15.32 (15.46)	- -	29.23 (29.55)
$[Ni_2(BAAD)_2](ClO_4)_4$	Light green	19.01 (19.09)	- -	18.02 (18.50)
$[Co_2(BAAD)_2Cl_4]$	Orange	17.93 (18.09)	21.7 (21.8)	25.6 (26.70)
$[Co_2(BAAD)_2Br_4]$	Reddishbrown	14.1 (14.21)	39.2 (38.5)	19.9 (20.03)
$[Co_2(BAAD)_2(NO_3)_4]$	Orange red	15.28 (15.56)	- -	29.4 (29.55)
$[Cu_2(BAAD)_2]Cl_4$	Dull green	18.97 (19.21)	21.21 (21.48)	25.23 (25.41)
$[Cu_2(BAAD)_2]Br_4$	Dark brown	14.92 (15.13)	38.98 (39.14)	19.87 (20.02)
$[Cu_2(BAAD)_2](ClO_4)_4$	Yellowish green	13.62 (13.84)	- -	18.21 (18.32)

BAAD= $C_8H_{16}N_6$,*Including nitrate nitrogen.

Infrared spectra:

The IR spectra of metal complexes have been studied by comparing the ligands 3,6- dimethyl-2,7-dioxo- 4,5-diazaocta-3,5-diene-2,7-dihydrazone and other related compounds in the region 4000- 400cm^{-1}.Some of the important and significant bands have been presented in Table-2. All the spectra show similar characteristics suggesting their iso structural nature. Many of the important bands in these spectra indicate the mode of bonding of the ligand with the metal ion. The spectral bands are in accordance with the structures of the complexes displayed in Fig- 1 and Fig-2. The ligand exhibits two characteristic bands at around 3330 cm^{-1} and 3125cm^{-1} with medium intensity arising out of $\upsilon_{asymm}(-NH_2)$, $\upsilon_{sym}(-NH_2)$ arising out of the terminal $-NH_2$ groups. The position of these bands is lowered to some extent than normal $\upsilon(NH_2)$. They occupy an intermediate position compared to 1,2,- diphenylethane-1,2- dione dihydrazone [13] and 2,3- butanedione dihydrazone [14]. Such a shifting is attributed to the change in vibrational energy for $-NH_2$ group due to the high degree of conjugation. The concerted shifting of the two bands to lower frequency region in all the metal complexes is in accordance with already reported dihydrazonecomplexes [7].

Apart from these structurally important bands, a series of three to four bands are observed in the region 3100-2700 cm^{-1} which are weak in intensity and can be assigned unequivocally to υC-H vibrations [18].

In the finger print region, two bands which are of considerable structural significance are observed ~1645 cm^{-1} and ~1565 cm^{-1} for the free ligand. The former is relatively broad and of medium intensity whereas the latter appears as a sharp and strong band. The high frequency band is least perturbed in almost all the metal complexes, while the low energy band moves towards a higher frequency region by 15 to 30 cm^{-1} with increased intensity. Taking into consideration their behaviour and relative intensities, the former band is attributed to δ NH_2 vibrations and the latter to $\upsilon(C \equiv N)$. There is no other band comparable in sharpness and intensity than the υ(C=N) in this region implying that all the azomethine linkages have similar vibration energies. The shift of υ(C=N) to a higher frequency region in the metal complexes clearly demonstrates that the linkage of the ligand with the metal ions occurs in a bi-bidentate manner with coordination taking place through azomethine nitrogen atoms. Far I.R. spectra, electronic spectra, magnetic susceptibility and thermochemical data corroborate these structures.

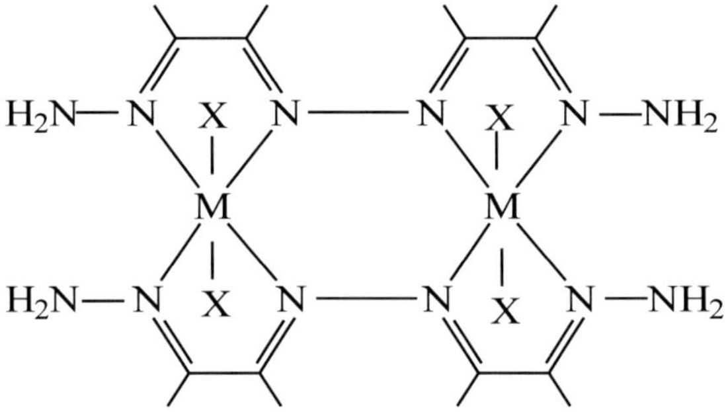

Fig-1M= Co(II) and Ni(II), X= Cl^-, Br^-, NO_3^-

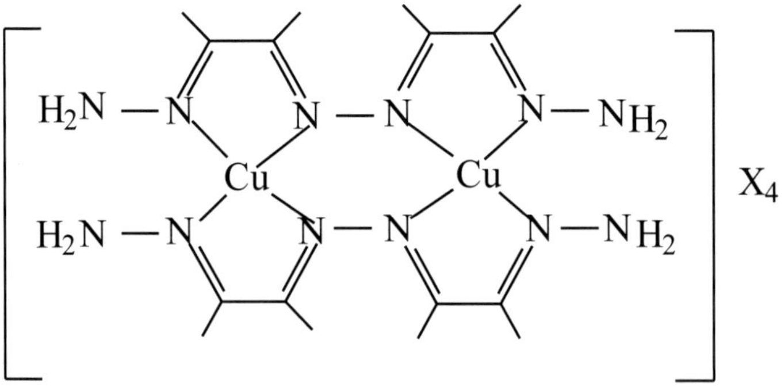

Fig-2X= Cl^-,Br^-,NO_3^-, ClO_4^-

All the complexes exhibit υ(M-N) in the range 530-570 cm^{-1}. For the Ni(II) and Co(II) perchlorate complexes a broad band of strong intensity is observed at ~1050-1120 cm^{-1} which is characteristics of uncoordinated ClO_4^- ion [19] and confirm them to be present in the outer-sphere of the complex ions. The nitrate complexes of Co(II) show three distinct bands around 1080, 1260 and 1415 cm^{-1} and suggest the nitrate groups to be coordinated in a unidentate manner [20]. Absence of these bands and higher molar conductance value of Copper(II) complexes suggest them to be of ionic character. A series of bands for the metal complexes as well as the ligand are observed around 1460, 1400, 1310, 1215, 915 and 755 cm^{-1}.They show closely similar features and positions in all the complexes and originate due to deformation vibrations.

Table-2 Structurally important infrared bands of BAAD and its metal complexes (in cm^{-1})

Complexes	$υ_{as}(NH_2)$ $υ_s(NH_2)$	υ(C-H)	υ(C=N)	$δNH_2$	υ(M-N)	υ(N-O) (NO_3)	δ CH_3
BAAD	3330s* 3125s	2850w	1565s	1645m	-	-	1370m 1300m
$[Ni_2(BAAD)_2Cl_4]$	3300s 3135s	2980w 2910w	1580s	1650m	538s	-	1455m 1400m 1390m
$[Ni_2(BAAD)_2Br_4]$	3259s 3145s	2020w 2970w	1585s	1645m	535s	-	1445m 1415m 1375m
$[Ni_2(BAAD)_2(NO_3)_4]$	3310s 3162s	2960w 3000w	1580vs	1645m	530s	1390s	1405m 1380m
$[Ni_2(BAAD)_2](ClO_4)_4$	3310s 3130s	2900w	1590s	1640m	535s	-	1455m 1400m
$[Co_2(BAAD)_2Cl_4]$	3335s 3170s	3000w 2930w	1595s	1650m	542s	-	1445m 1420m 1385m
$[Co_2(BAAD)_2Br_4]$	3320s 3170s	3030w 2980w	1575s	1645m	554s	-	1445m 1400m 1385m

$[Co_2(BAAD)_2(NO_3)_4]$	3320s 3162s	2965w	1570s	1640m	534s	1415s 1260s 1080s	1395m 1305m
$[Co_2(BAAD)_2](ClO_4)_4$	3330s 3178s	2925w	1580s	1645m	535s	-	1455m 1422m 1395m
$[Cu_2(BAAD)_2]Br_4$	3300s 3160s	2915w	1585s	1635m	537s	-	1460m 1420m 1380m
$[Cu_2(BAAD)_2]Cl_4$	3260s 3170s	2910w	1575s	1645m	558s	-	1445m 1430m 1380m
$[Cu_2(BAAD)_2](ClO_4)_4$	3300s 3190s	292w	1586s	1645m	536s	-	1460m 1405m 1350m

BAAD= $C_8H_{16}N_6$, s=strong, vs= very strong, m=medium, w=weak

Electronic spectra and magnetic properties:

Nickel(II) complexe*s*:

The electronic spectra for the present series of complexes have been studied in the region 6000 to 30000 cm^{-1}. The possible assignments and spectral bands are recorded in Table- 3 and two representative spectra are shown in Fig-3.

The spectra show three distinct bands and clearly approach the spectral features of octahedral Nickel(II) complexes. The broad band in the range 9000 to 11,000 cm^{-1} and two intense bands around 21,000 and 29,000 cm^{-1} are assigned under octahedral micro asymmetry to $^3A_{2g}\rightarrow{}^3T_{2g}$, $^3A_{2g}\rightarrow{}^3T_{1g}(F)$ and $^3A_{2g}\rightarrow{}^3T_{1g}(P)$ transitions respectively. The high intense third band arises due to charge transfer contributions. The width of the first band manifests distortion of the complex most probably along the trigonal axis. Trigonal distortion is uncommon for Nickel(II) complexes. However some trigonal molecules are known [21,22] and their electronic spectra have been characterized [23]. They closely resemble spectral features of octahedral complexes.

Apart from the aforementioned transitions all of these complexes show a shoulder near 14,000 cm^{-1} which is attributed to $^3A_{2g}\rightarrow{}^1T_{2g}$ (^1D) spin forbidden transition.

Table-3 Electronic spectral bands (in cm^{-1}) of Nickel(II) complexes with BAAD.

Complexes	μ_{eff} in B.M.	$^3A_{2g}\rightarrow{}^3T_{2g}$ (cm^{-1})	$^3A_{2g}\rightarrow{}^1T_{2g}(^1D)$ (cm^{-1})	$^3A_{2g}\rightarrow{}^3T_{1g}(F)$ (cm^{-1})	$^3A_{2g}\rightarrow{}^3T_{1g}(P)$ +CT (cm^{-1})
$[Ni_2(BAAD)_2Cl_4]$	2.72	9,820	14,100sh	23,500	29,400
$[Ni_2(BAAD)_2Br_4]$	2.93	9,700	14,300	21,000	29,550
$[Ni_2(BAAD)_2(NO_3)_4]$	2.99	-	14,400	22,000	28,870

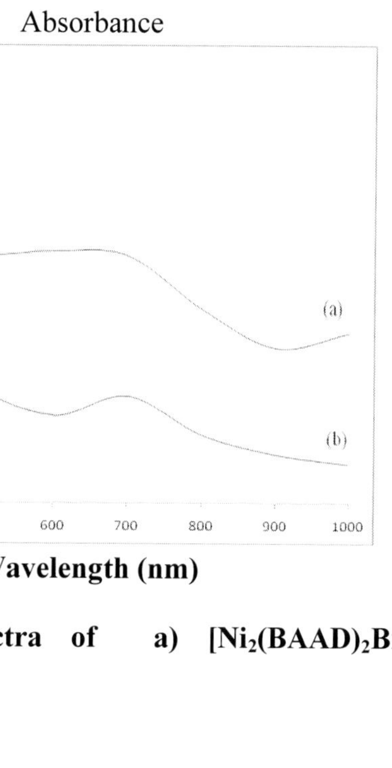

Fig-3 Electronic spectra of a) $[Ni_2(BAAD)_2Br_4]$ b) $[Ni_2(BAAD)_2(NO_3)_4]$

Copper (II) complexes :

The Copper (II) complexes show subnormal magnetic moments in the range 1.30 to 1.61 B.M. per copper atom having the ground term ^2D. The low magnetic moments are attributable to possible intermolecular or intramolecular magnetic interactions between Copper(II) ions. The band positions and assignments are recorded in Table-4 and a representative spectrum is shown in Fig-4. They show an unsymmetric and broad ligand field band around 15,100 cm^{-1} except the perchlorate complex for which the ligand field band is located ~19,200 cm^{-1}. The width and asymmetry manifest two or three superimposed bands comparable both in magnitude and width with established Cu(II) complexes having similar low symmetry ligand fields with CuN_4 chromophores [24,25].

Table-4 Electronic spectra of Copper (II) complexes with BAAD.

Complexes	μ_{eff} in B.M.	Ligand field band in cm^{-1}	Remarks
$[Cu_2(BAAD)_2]Cl_4$	1.60	15,200	
$[Cu_2(BAAD)_2]Br_4$	1.31	15,150	Broad and asymmetric
$[Cu_2(BAAD)_2](ClO_4)_4$	1.39	19,200	

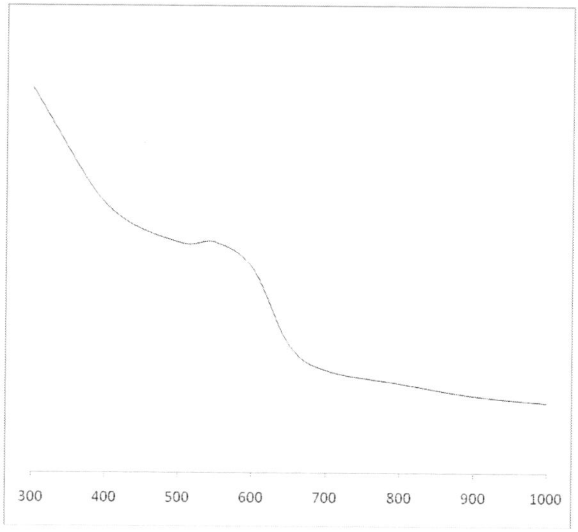

Fig- 4 Electronic Spectra of $[Cu_2(BAAD)_2](ClO_4)_4$

Cobalt(II) complexes :

The Cobalt(II) complexes are all paramagnetic and their magnetic moments lie in the range 4.1 to 4.7 B.M. per cobalt atom. The electronic spectra of the complexes have been measured in the range 6500 to 30,500 cm^{-1}.

The electronic spectra (Table-5 and Fig-5) show ligand field transition for chloro, bromo complexes at ~16,200 cm^{-1} attributable to $^4T_{1g}(F) \rightarrow ^4A_{2g}(F)$ transition. However this band is not observed in case of the nitrato complex due to the overlapping of the 20,900cm^{-1} band. The next important band for all the complexes is located in the region 19,000-21,000 cm^{-1} which arises due to $^4T_{1g}(F) \rightarrow ^4T_{1g}(P)$ transitions. These two important bands generally characterize octahedral or more appropriately to tetragonal ligand field about the metal ions [26]. In addition to the above bands all the cobalt complexes exhibit a strong CT band beyond 25,500 cm^{-1}.

Table-5 Electronic spectra of Cobalt (II) complexes with BAAD (cm^{-1}).

Complexes	μ_{eff} in B. M.	$^4T_{1g}(F)\rightarrow{}^4A_{2g}$ (F)	$^4T_{1g}(F)\rightarrow{}^4T_{1g}$ (P)	C.T. Band
$[Co_2(BAAD)_2Cl_4]$	4.7	16,300	19,100	27,500
$[Co_2(BAAD)_2Br_4]$	4.2	16,000	20,000	28,900
$[Co_2(BAAD)_2(NO_3)_4]$	4.1	-	20,900	31,000
$[Co_2(BAAD)_2](ClO_4)_4$	4.3	15,900	21,200	28,700

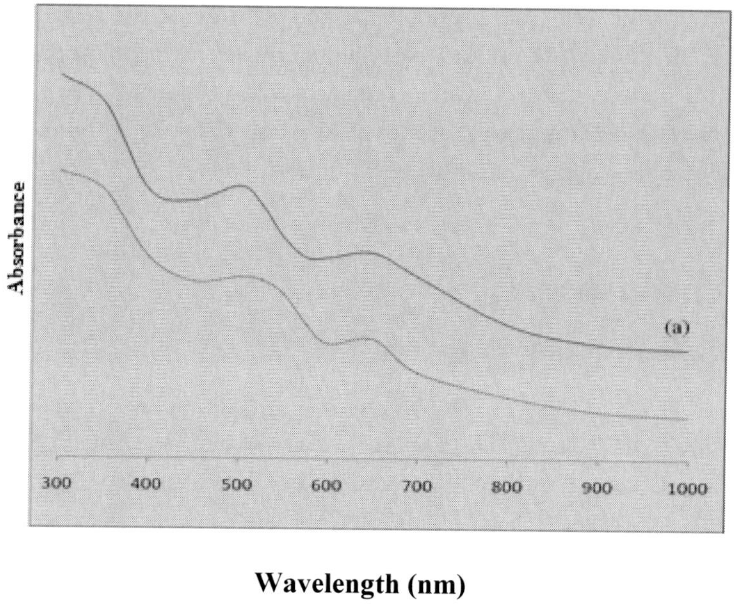

Fig-5 Electronic Spectra of a) $[Co_2(BAAD)_2Cl_4]$ b)$[Co_2(BAAD)_2Br_4]$

Thermal analysis:

Thermogravimetric and differential thermal analyses of the complexes have been carried out from the ambient temperature of the furnace to 1000^0 C. Thermal behaviour of the complexes follows a complex pattern and varies from compound to compound showing thermal stability in varying degrees. Almost all the complexes show three major decomposition temperature ranges 400-500^{0}C. 500-600^{0}C and 700-800^{0}C and are marked with exothermic peaks in DTA curves implying dissociation of the ligand with synchronous oxidation to simpler products. The complexes containing nitrate ions on the other hand are found to be thermally least stable and begin to decompose at a relatively low temperature. For all the compounds the final products are found to be of the respective metal oxide. Attempts to correlate stepwise ligand dissociation with temperature were unsuccessful. **CAUTION:** The perchlorate complexes are explosive in nature and violently decompose above 320-330 ^{0}C.

EXPERIMENTAL

Materials and Methods:

2, 3- butanedione, hydrazine hydrate, dimethylformamide(DMF) and DMSO were of BDH quality and used as supplied. Perchloric acid was of Fluka quality. Bromide salts of copper(II), nickel(II) and cobalt(II) were prepared by dissolving the respective metal carbonates in minimum volume of hydrobromic acid follwed by crystallization. For the preparation of cupric bromide it was essential to add a few drops of bromine to get the pure product [27].

Preparation of 3,6-dimethyl-2,7-dioxo-4,5-diazaocta-3,5-diene-2,7-dihydrazone (BAAD)-($C_8H_{16}N_6$):

The ligand BAAD was synthesized with a slight modification of the literature method[4]. To an alcoholic solution of 2,3-butanedione (17.2 g, 0.2 mol) at 0^{0}C was added an alcoholic solution of hydrazine hydrate (5 g, 0.1 mol) with stirring over a period of 30 mins. The resulting solution was kept at 0^{0}C for 12-20 hrs. and was again added slowly to a solution of hydrazine hydrate (10 g, 0.2 mol). The yellow solution was allowed to crystallize by slow evaporation at room temperature and finally recrystallized from ethanol. The product was obtained as a bright yellow solid. Yield 60% [melting point 130^0

- 135^0C (lit.melting point 1270^0 - 1410^0 C) found N, 42.2%, calc.N, 42.855].

Preparation of tetrachloro- bis (3,6-dimethyl-2,7-dioxo-4,5-diazaocta-3,5-diene-2,7 dihydrazone)dicobalt(II) [$Co_2(C_8H_{16}N_6)Cl_4$]:

About 4 g of BAAD (0.02 mol) was dissolved in 25 ml of absolute ethyl alcohol by gently warming below 70^0 C. To this warm solution an alcoholic solution of $CoCl_2$.$6H_2O$ (4.5 g, 0.03 mol) was added drop wise with stirring. A pink red product was formed after few minutes. It was stirred for thirty minutes, allowed to settle and quickly filtered. It was repeatedly washed with absolute alcohol, ether and dried in vacuo.

Bis-bromo, nitrato and perchloratoCo (II) complexes were prepared following similar procedures.

Copper(II) and Nickel(II) complexes were synthesized following similar method as the Cobalt (II) complexes.

PHYSICAL MEASUREMENTS:

The complexes have been characterized on the basis of different physicochemical analysis.

Infrared spectra of the complexes were recorded in KBr phase using Shimadzu-FTIR-121 Spectrophotometer.nThe electronic spectra of the compounds were recorded in nujol mull in a Shimadzu-2450 UV-Visible Spectrophotometer. Magnetic susceptiblity of the compounds were measured by Gouy balance using Hg_2 [$Co(CNS)_4$] as the calibrant. The thermoanalytical measurements TG and DTA were done with a NETZSCH Geratebau. GmbH thermal analyzer.

ANTIFUNGAL ACTIVITY:

Microorganisms tested:

The microbial strains (*T. viridae, Aspergillusflavus,* and *Aspergillusniger*) were obtained from National Chemical Laboratory (NCL), Pune. The fungi were maintained on nutrient agar at 4°C and sub cultured before use. The microorganisms studied are clinically important ones causing several infections, food borne diseases, spoilages, skin infection and it is essential to overcome them through some active therapeutic agents.

Method:

In the present investigation standard potato dextrose agar medium was used for antifungal activity by well diffusion method[28]. DMSO was used as the solvent and Fluconazole (antifungal agent) as control. The PDA (potato dextrose agar) medium was prepared and inoculation was done inside the Laminar Air Flow. A well was made on the agar medium inoculated with microorganisms. The well was filled with the 50 μltest solution, covered with petri plates and they were incubated at 35°C for 72 hrs. During the incubation period, the solution was diffused and affected the growth of the inoculated microorganisms. The inhibition zone was developed and values of the compounds against the growth of microorganisms are recorded (Table-.6).

Statistical Analysis:

Antifungal data were analyzed by one-way Analysis of Variance (ANOVA) followed by Dennett's t *test* with equal sample size ($n = 3$). The difference was considered significant when p value < 0.05. All the values were expressed as mean ± standard error mean (S.E.M.)

DISCUSSION

The inhibitory activity of metal complex was observed in fungal pathogens. A comparative study of the ligand and its complexes (Zone of inhibition) indicates that all of the metal chelates show higher antibacterial activity than the free ligand. The increased activity of the metal chelates is due to increased lipophilicity of the metal ions. Among the three metal complexes, Cu metal complex was highly susceptible than others complexes. They showed activity in Co < Ni < Cu manner. Such increased activity of the complexes can be explained on the basis of Overtone's concept[29] and Tweedy's Chelation theory[30]. According to Overtone's concept of cell permeability, the lipid membrane that surrounds the cell favours the passage of only the lipid-soluble materials due to which liposolubility is an important factor, which controls the antifungal activity. On chelation, the polarity of the metal ion will be reduced to a greater extent due to the overlap of the ligand orbital and partial sharing of the positive charge of the metal ion with donor groups. Further, it increases the delocalization of π-electrons over the whole chelate ring and enhances the lipophilicity of the complexes. The increased lipophilicity enhances the penetration of the complexes into lipid membranes and blocking of the metal binding sites in the enzymes of microorganisms. These complexes also disturb the respiration process of the cell and thus block the synthesis of the proteins that restricts further growth of the organism. The antifungal activities of the complexes are recorded in Table-6, Fig-6.

Table-6 Antifungal activity of synthesized complexes.

Sl.No	Compound	*A.niger*	*A. flavus*	*T. viridae*
1	BAAD	1.15±0.067	1.09±0.173	1.10±0.231
2	$[Co_2(BAAD)_2Cl_4]$	1.28±0.133	1.21±0.219	1.25±0.252
3	$Ni_2(BAAD)_2Cl_4]$	1.50±0.120	1.40±0.176	1.43±0.153
4	$[Cu_2(BAAD)_2]Cl_4$	2.09±0.088	2.01±0.260	2.05±0.346
5	Fluconazole	2.80±0.346	2.70±0.176	2.73±0.153

Values are expressed as Mean±S.E.M. n=3

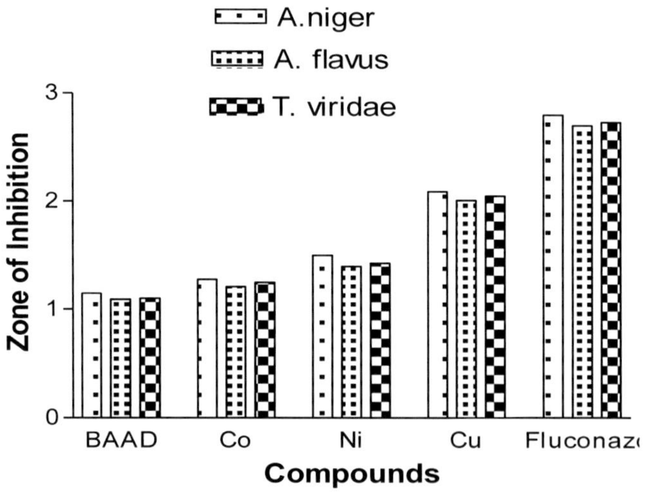

Fig-6 Antifungal activities of synthesized compounds against selected fungus.

CONCLUSION

The antifungal activity of the binuclear macrocyclic complexes is further enhanced due to association with a metal ion which enhances lipophilicity. The open chain diacetyldihydrazone (BAAD) metal complexes have been prepared as precursors for the preparation of macrocyclic complexes. In this investigation the Zone of Inhibition of the open chain ligand (BAAD) and its metal complexes along with the macrocyclic complexes (with ODDO, OMDO, ODTO) are tested and their order of fungitoxicity are correlated. All the metal complexes have been screened with fungal pathogens namely (*A.niger, A. flavus,* and *T. viridae*) against a standard antifungal agent Fluconazole.

Smart materials, be they of physical or biological origin have many applications: *Bacteriorhodopsin (BR) thin film*-based spatial light modulator (SLM) and erbium/ytterbium (Er^{3+}+Yb^{3+}) doped *polymethylmethacrylate* (PMMA) -crown glass optical waveguides for the study of retinal response; the pulse laser deposited **high dielectric oxides** in device fabrication; **pseudogap-induced high temperature superconductivity** for specific heat anomaly in cuprates and tunneling spectra; the robust antifungal properties of new generation **binuclear macrocyclic complexes**; antidiarrhoeal and the anti ulcer activity of the dietary fiber of ***C. fistula and T. indica*;** the antidiabetic, antihypertensive, diuretic and antioxidant properties of **ScopariadulcisLinn** whose administration has shown significant reduction in systolic blood pressure on DOCA induced hypertensive rats; and the ethanol extract of the ***roots of SidaacutaBurm*** F. (Malvaceae) exhibiting potent post coital, anti-implantation, abortion, estrogenic and antiestrogenic activity and reduction in sperm motility.

REFERENCES

[1] D.H. Busch, *Acc. Chem. Res.*, (1978), 11, 392.

[2] G.A. Melson, in: G.A. Melson (Ed.), *Coordination Chemistry of Macrocyclic Compounds*, Melson, Plenum Press, New York, (1979), p- 2.

[3] D. P. Singh, R. Kumar, V. Malik and J. Singh, *J. Enzyme Inhib. Med. Chem.*, (2009),**24(**5), 1201.

[4] S. Ilhan and H. Temel, *Trans. Met. Chem.*, (2007),**32(8),** 1039-1046.

[5] M. Shakir, S. Khatoon, S. Parveen and Y. Azim, *Trans. Met. Chem.*, (2007), 32(1), 42.

[6] S. Chandra and M. Pundir, *Spectrochim. Acta.*, (2008),**69**, 1.

[7] D. P. Singh, R. Kumar, V. Malik and P. Tyagi, *J. Enzyme Inhib. Med. Chem.*, (2007), **22**(2), 177.

[8] R.P. Lindstead and A.R.Lowe, *J.Chem. Soc.*,(1934), 1022-1027.

[9] M. F. Iskander and S. Saddeck, *Inorg. Chim. Acta*, (1977), 22, 141.

[10] M. F. Elshazly, *Transition Met. Chem.*, (1981), 6, 8.

[11] M. F. Iskander, L. Elsayeed and S. Saddeck, *Transition Met. Chem.*, (1981), 6, 32.

[12] R. C. Stoufer and D. H. Busch, *J. Am. Chem. Soc.*, (1960), 82, 4834.

[13] J. Chakrabarty and B. Sahoo,*Indian J. Chem.*, (1980), 19A, 441.

[14] R. C. Stoufer and D. H. Busch, *J. Am. Chem. Soc.*, (1960), 82, 3491.

[15] H. C. Rai and B.Sahoo, *J. Indian Chem. Soc.*, (1976), 53, 646.

[16] V. L. Goedken and K.P. Weinwright, *J. Chem. Soc., Chem. Comm.*, (1973), 62.

[17] V. L. Goedken, '*Inorganic Syntheses*' Vol.XX ed. D.H. Busch, (1980),87.

[18] L. J. Belamy, 'The Infrared Spectra of complex molecules' (Wiley,New York), (1973), p-17.

[19] K. Nakamato, *Infrared Spectra and Raman Spectra of Inorganic and Coordination Compounds* (Wiley,Interscience,New York, Brisbane,Toronto), (1978), p-227.

[20] P. W. Ball and A. B. Blake, *J. Chem. Soc. (A)*, (1969), 1415.

[21] J. E. Parks, B.E. Wagner and R. H. Holm, *J. Am. Chem. Soc.*, (1970), 92, 3500.

[22] J. E. Parks, B. E. Wagner and R. H. Holm, *Inorg. Chem.*, (1971), 10, 2472.

[23] E. Larsen, G. N. Lamar, B. E. Wagner, J. E. Parks and R. H. Holm, *Inorg. Chem.*, (1972), 11, 2652.

[24] A .Chakravorty and T. S. Kannan, *J. Inorg. Nucl. Chem.*, (1967), 29, 1691.

[25] R. E. Clarke and J. H. Weber, *J. Inorg. Nucl. Chem.*, (1968), 30, 1837.

[26] J. Fergison, *Prog. Inorg. Chem.*, (1970), 12, 249.

[27] J. Chakrabarty and B. Sahoo, *Indian J. Chem.*, (1981), 20A, 431.

[28] O. N. Irobi, M. Moo-Young and W. A. Anderson, *Int. J. Pharm.*, (1996), 34, 87-90.

[29] Y. Anjaneyula and R. P. Rao, *Synth. React. Inorg. Met .Org. Chem.*, (1986), 16, 257-272.

[30] R. S. Srivastava ,*Inorg. Chim. Acta.*, (1981), 56, L65-L67.

United Scholars
Publications

Email: info@unitedscholars.net
www.unitedscholars.net

Made in United States
North Haven, CT
04 November 2026

21764586R00128